SOLUTIONS OF ORDINARY LINEAR DIFFERENTIAL EQUATIONS WITH CONSTANT COEFFICIENTS (OLDECC)

Solutions of Ordinary Linear Differential Equations With Constant Coefficients (OLDECC)

EVERARD M. WILLIAMS

Professor of Electrical Engineering
Carnegie-Mellon University
Pittsburgh, Pennsylvania

ASOK K. MUKHOPADHYAY
Formerly Senior Research Engineer
Jet Propulsion Laboratory
Pasadena, California

John Wiley & Sons, Inc.
New York • London • Sydney • Toronto

The United States Government may use, reproduce publish, or have reproduced,
used, and published, without charge for United States governmental purposes, all
or any part of the book in accordance with the terms of National Science Founda-
tion grant GR–2441 under which this work was prepared.

Library of Congress Catalog Card Number: 68–26854
SBN 471 94730 X
Printed in the United States of America

PREFACE

To the Instructor

This programmed work is intended for use by students for self-instruction. It should not require classroom reinforcement, except possibly in a "final" examination, which will serve to motivate them to study the text and to evaluate their success in this study.

The subject treated is the solution of ordinary linear differential equations with constant coefficients (abbreviated OLDECC for convenience). It is assumed that many of the students who use this programmed material will have covered the subject very briefly in an earlier or concurrent mathematics course. In most widely used texts about two pages are devoted to the general treatment of linear differential equations with constant coefficients, three pages to the solution of second-order homogeneous equations, about two and one-half pages to second-order inhomogeneous equations, and five pages to one example solution of an oscillating spring-mass problem. From the mathematics instruction point of view this brief treatment is more than adequate; what is needed in addition, however, by students in engineering and physical sciences is enough exercise of the mathematical methods to develop solid student competence. This competence is a *sine qua non* for much work in these fields.

Although this book contains no material not included in other mathematics texts, the emphasis is different because the concentration is on a few basic concepts and extensive student practice examples are provided, which are criticized and corrected at each step without the intervention of the instructor. It should be possible to use them to eliminate much tedious class work and instructor-graded homework.

This programmed text may definitely be used by students who have not studied differential equations in a previous mathematics course. The minimum mathematical preparation (in calculus) required is an understanding of and skill in differentiation and familiarity with the concept of integration. A pretest of minimum prerequisites may be found on page *xi*. In our opinion, however, the work should be undertaken only by students who need and will use the differential-equation solution skill almost immediately in one or more physical science, engineering science, or engineering courses. Otherwise, as experience shows, students lose the skills that are learned rather rapidly.

The minimum time required for an average student to complete this work, if a number of the more elaborate practice examples are omitted, is about five hours. If every example is done, the time required is closer to 10 hours.

This book is one of a group of programmed texts which have been used experimentally at Carnegie-Mellon University in a basic course in electric-circuit theory. Some of the results obtained have been reported elsewhere,* but, it has not been practicable to separate from the overall educational results the contributions of the OLDECC program alone. This program, however, has clearly fulfilled its function and its use is continuing.

Behavioral Objectives

Before outlining the behavioral objectives for this work, we wish to emphasize that they do not include teaching students to set up differential equation models for physical problem configurations. This programmed text deals with finding solutions to differential equations that have already been set up. Although in numerous examples the student starts with the problem and goes through the process of setting up the equation (and therefore acquires or sharpens some "setting up" skills), he is taken through this process primarily so that he can learn the critical significance in the solution of and how to use initial conditions and the relation of complementary and particular solutions to the transient and steady-state solutions.

The examples include a number of electric-circuit problems as well as mechanics problems, that are organized so that a student who is not familiar with electric-circuit theory may "skip" the setting up of the differential equations and go directly to the solution process, which does not require electric-circuit theory.

After completion of Chapter I the student should be able to identify clearly the differential equations that fall into the category of ordinary linear differential equations with constant coefficients (OLDECC). To do this he should be able to identify a *differential equation*, determine its *order* and *degree*, and distinguish between *independent* and *dependent* variables, *linear* and *nonlinear* differential equations, *constant* and *variable coefficients* in a differential equation, *ordinary* and *partial* differential equations, and *homogeneous* and *inhomogeneous* differential equations. The student should be able to find the mathematical solution of the homogeneous OLDECC, define the relation of the undetermined (ini-

---

*"Innovation in Undergraduate Teaching," E. M. Williams, *Science,* **155**, *3765*, 974—979 (February 24, 1967).

tial condition) coefficients to the configuration of the physical problem
from which the differential equation arises, and, given the initial conditions, find the actual values of the coefficients. In the process the
student will have learned to define and use the new terms: *auxiliary
equation, undetermined coefficients, initial conditions, damping force,
underdamped, critically damped,* and *overdamped.*

After completion of Chapter II the student should be able to find the
complete solution of the inhomogeneous OLDECC with a steady or sinusoidal driving function and define the relation of the particular and complementary solutions to the steady-state and transient behavior in the
physical configuration from which the differential equation arises. In
the process of acquiring these skills the student will have learned to
define and use the terms: *driving function, forcing function, driving
force, particular solution, particular integral, complementary solution,*
and *complementary function.*

After completion of Chapter III the student should be able to find the
complete solution of the inhomogeneous OLDECC with polynomial and
exponential driving functions.

No new principles, concepts, or skills are acquired by completing
Chapter IV; the purpose of this portion of the program is to enable the
student to *improve* his ability to use the skills, terms, and concepts of
the first three chapters and to relate them to what he may have learned
elsewhere concerning physical configurations and differential equation
modeling of these configurations.

To the Student

The purpose of this book is to guide you in learning by yourself to
solve a common class of differential equations—ordinary linear differential equations with constant coefficients (OLDECC).

This work requires some definite prerequisite mathematical skills
that involve both algebra and calculus. You must have some skill in
differentiation and be familiar with the concept of integration. To assist
you in determining whether you have the necessary prerequisites, a test
of the minimum skills required appears on page *xi*. If you find any of
the test questions difficult to answer, you should review corresponding
material in an algebra or calculus text or course. Except for some review of this kind, however, this programmed work needs no reference to
other works; it is self-contained.

In order to use this book to obtain the solutions of ordinary linear differential equations with constant coefficients, you do not need to have
taken a course in mathematics (or another type of course) that treats

such solutions. Even if you have done so, our experience with students who use this text shows that it is a fruitful supplement and will provide you with the solid competence that your work in courses in science and/ or engineering will demand.

This is a programmed text. If you follow instructions carefully, you will find that learning with this work is like having a private tutor. You must follow the instructions explicitly, however. This book is not one that you can "leaf through" to find what you want, skipping other matters not of immediate interest. *There is an index that will probably be valuable for review purposes only.* We can assure you that a careful step-by-step first reading will prove worthwhile.

Before starting your study, you should read the instructions that follow carefully.

Instructions for the Use of This Book

This is a programmed self-instructional text. It is divided into a large number of small sections, called frames, and is designed to help you learn by the methods of programmed learning; when you reach any particular frame, you will study it and then have an opportunity to respond to some implicit or explicit question concerning the material in it. You will thus check and strengthen your understanding of each small portion of the subject before proceeding. By dividing the subject into small portions and reinforcing your understanding of each of them, step by step, it is possible for the writer of a programmed work to encourage more thorough comprehension of the subject matter than is possible with most ordinary books or even in ordinary classroom teaching.

The programmed text asks you to respond to questions in several different ways—typically by filling in a blank, indicating "true" or "false," "yes" or "no," or selecting one or more responses from a group of several (multiple-choice) answers.

In most cases, after you have responded to a question, the correct answer is immediately available to you to compare with your own. A typical frame is described to show you how this is done.

---

1. In a number of computer programming languages the symbol * (asterisk) indicates the operation of multiplication; for example, with this notation the result of the operation indicated as 2 * 4 would be

2 * 4 = [_____].

- - - - - - - - - - - - - - - - -

2 * 4 = 2 × 4 = 8

This is a complete frame. It first makes a statement to teach you the meaning of the asterisk in a programming language. It then asks you a question that will make sure that you have understood the statement; to answer this question you fill in the indicated blank. Finally, the correct response appears on the line below your response so that you can check its accuracy. Immediately below this reinforcement is a rule which indicates that the frame is complete and that the next frame will follow.

What are you to do if a response that you have given proves to be incorrect? In a simple case the reinforcement should be sufficient to show how the correct response is obtained and to enable you to find your mistake. In others the situation may be more complicated and you may not be able to uncover the difficulty. In most of these cases you will find that explanations are given, often arranged so that those who have made a correct response can pass it by. A typical instruction states, ''If your response to the above question was correct and you are satisfied that you understand the frame, skip the following explanation and go directly to Frame 9 on page 11.''

If you read each page as you would an ordinary book, you would probably see the correct response before you had a chance to construct it. It is not really possible for you to ''close your eyes'' to the reinforcements given and concentrate on your own responses. To avoid this difficulty you should use the mask provided. This mask exposes one line at a time as it is moved down the page. When you reach a response (blank, multiple choice, etc.), the mask will cover the correct answer while you are constructing your response. You then slide the mask down to uncover the correct response, which you can use to check your own work.

Once in a while you may be unable to construct a response because the frame has not been entirely clear to you. You certainly may look at the reinforcement for help if you are bogged down at any time, but you should make a genuine effort to construct your own responses as often as possible. The object is to learn, not to speed through the work.

EVERARD M. WILLIAMS
ASOK K. MUKHOPADHYAY

Pittsburgh, Pennsylvania
Pasadena, California
June 1968

TEST OF MINIMUM MATHEMATICAL
PREREQUISITES REQUIRED

1. Differentiate the following functions, as indicated:

(a) $\dfrac{d}{dx}(1 + 3.4x + 3.1x^2 + 0.01x^3 - 0.1x^4)$

$= [\underline{\hspace{8cm}}]$

(b) $\dfrac{d}{dx}[f(x)]$ where $f(x) = a + bx + cx^3 - dx^5$

$= [\underline{\hspace{8cm}}]$

(c) $\dfrac{d}{dt}[f(t)]$ where $f(t) = a \sin \omega_1 t + b \cos \omega_2 t + gt \sin \omega_3 t$

$= [\underline{\hspace{8cm}}]$

(d) $\dfrac{d}{dy}(e^{ky}) = [\underline{\hspace{7cm}}]$

2. Find the roots of the following algebraic equations:

(a) $3.2m + 4 = 0$ $m = [\underline{\hspace{4cm}}]$

(b) $m^2 - 4m + 3 = 0$ root(s) are $[\underline{\hspace{4cm}}]$

(c) $m^2 - 6m + 9 = 0$ root(s) are $[\underline{\hspace{4cm}}]$

(d) $m^2 - 2m + 2$ root(s) are $[\underline{\hspace{4.5cm}}]$

(e) $m^3 - 6m^2 + 11m - 6$ one root is $m = 1$, the others are
$[\underline{\hspace{4cm}}]$

3. Given the following equations, find the indicated numerical values of the coefficients A, B, C, D, etc.:

(a) $A \cos \omega_1 t + B \sin \omega_1 t = 3.4 \cos \omega_1 t + 4.1 \sin \omega_1 t$
$A = [\underline{\hspace{3cm}}]$ $B = [\underline{\hspace{3cm}}]$

(b) $Ae^{-3t} + Bte^{-3t} = 1.1e^{-3t} + 0.2te^{-3t}$
$A = [\underline{\hspace{3cm}}]$ $B = [\underline{\hspace{3cm}}]$

3. (c) In the equation

$$0.01 \frac{d^2x}{dt^2} + 0.1 \frac{dx}{dt} + x = 5 \cos 10t - 10 \sin 10t$$

it is known that $x = A \cos 10t + B \sin 10t$
The values of A and B, then, are
$A = [_____]$ $B = [_____]$

4. (a) What is the result of the differentiation with respect to time of the definite integral

$$f(t) = \int_0^t a\ dt$$

that is, what is

$$\frac{d}{dt} \int_0^t a\ dt\ ?\ \ [_____]$$

(b) Similarly

$$\frac{d}{dt} \int_0^t kt^2\ dt = [_____]$$

(c) Similarly, if i is a function of t,

$$\frac{d}{dt} \int_0^t \frac{i}{c}\ dt = [_____]$$

5. In the equation

$$i = e^{-5t}(A \sin 100t + B \cos 100t) + 15 \sin 72t - 12 \cos 72t$$

if, for $t = 0$, $i = 0$ and $di/dt = 120$, what are the necessary values of A and B?
$A = [_____]$ $B = [_____]$

6. If in the equation

$$\frac{d^2y}{dx^2} + 2\frac{dy}{dx} + 8y = 24\ e^{-4t}$$

the solution is stated to be of the form $y = z + 3$, then z must be the solution of the equation:

$$[_____]$$

7. (a) There are identities between the cosines and sines of real arguments and expressions containing exponential functions of imaginary arguments; for example, for $i = \sqrt{-1}$,

$$\cos x = [\underline{\hspace{5cm}}]$$

(b) $\quad \sin x = [\underline{\hspace{5cm}}]$

8. (a) The exponential expression $y = 5e^{-6ix}$, where $i = \sqrt{-1}$, can be rewritten in trigonometric form containing sines and cosines of real arguments. In this form

$$y = 5e^{-6ix} = [\underline{\hspace{7cm}}]$$

(b) The general expression for $x = Ae^{it}$ in terms of sines and cosines of real arguments is

$$x = Ae^{it} = [\underline{\hspace{6cm}}]$$

(c) The identity between an exponential with an imaginary exponent and an expression containing trigonometric functions of real arguments, for

$$i = \sqrt{-1}, \quad \text{is} \quad e^{ikt} = [\underline{\hspace{5cm}}]$$

At the completion of this test, turn to page 132 to check your answers.

CONTENTS

SOLUTIONS OF ORDINARY LINEAR
DIFFERENTIAL EQUATIONS
WITH CONSTANT COEFFICIENTS
(OLDECC)

Chapter I

HOMOGENEOUS ORDINARY LINEAR DIFFERENTIAL EQUATIONS WITH CONSTANT COEFFICIENTS

Part A—Some Elementary Examples

Section 1 TERMINOLOGY

In this section we will review some matters of terminology which you probably have already studied. Essentially, this short review is necessary so that we can be sure that the terms we will use mean exactly the same thing to both the student and the authors.

This terminology describes various types of differential equations.

---

1. First let us consider the differential equation

$$\frac{d^2y}{dt^2} + 6\,\frac{dy}{dt} + 5y = 0 \tag{1-1}$$

The Eq. (1-1) is called a differential equation because one or more of the terms in the equation are: [integers, fractions, derivatives].

— — — — — — — — — — — — — — — — — —

derivatives

Remark. This equation, incidentally, cannot be solved by a strictly algebraic process.

Any equation with terms in an unknown dependent variable or variables which are derivatives of the independent variable or variables will be termed a [_____] equation.

— — — — — — — — — — — — — — — — — —

differential

Reference Equation

1-1 $\dfrac{d^2y}{dt^2} + 6\dfrac{dy}{dt} + 5y = 0$

---

2. Since we are using the terms *independent* and *dependent* variable, let us make sure that we understand them.

In Eq. (1-1) the dependent variable is [y, t].

_ _ _ _ _ _ _ _ _ _ _ _ _ _ _ _ _ _ _

y

In the equation

$$\dfrac{d^4x}{dy^4} + 15\dfrac{d^3x}{dy^3} + 10\left(\dfrac{d^2x}{dy^2}\right)^2 + 0.3x = y\cos 377y$$

the independent variable is [x, y].

_ _ _ _ _ _ _ _ _ _ _ _ _ _ _ _ _ _

y

3. Since this work deals only with *linear* differential equations, we need to be able to distinguish between *linear* and *nonlinear* differential equations.

A linear differential equation is [_____

_____

_____].

- - - - - - - - - - - - - - - - - -

one in which neither the dependent variable, such as y in Eq. (1-1), nor any of the derivatives of this variable in the equation appear in any power other than the first.

Equation (1-1) is a linear equation because neither y, dy/dt or d^2y/dt^2 appears in any power other than the [_____].

- - - - - - - - - - - - - - - - - -

first

Indicate whether each of the following is a linear or nonlinear differential equation:

$$0.3 \frac{d^3y}{dt^3} + 3 \times 10^{-5} \frac{dy}{dt} + 10^{-7}y = 14t^2 \quad [\text{_____}]$$

- - - - - - - - - - - - - - - - - -

linear

$$15 \frac{d^2y}{dt^2} + 13.1 \frac{dy}{dt} + y^2 = 12t^2 \qquad [\text{_____}]$$

- - - - - - - - - - - - - - - - - -

nonlinear

$$0.3 \frac{d^2y}{dr^2} + \frac{15}{r} \frac{dy}{dr} + 3y = \sin kr \qquad [\text{_____}]$$

- - - - - - - - - - - - - - - - - -

linear

4. The terminology applied in defining a "linear" equation is also related to the matter of the *degree* of an equation. The degree of a differential equation is the [_____

_____

_____].

_ _ _ _ _ _ _ _ _ _ _ _ _ _ _ _ _ _ _

exponent to which the highest-order derivative is raised, after the equation has been cleared of fractions and radicals, if any such clearing is possible.

A linear differential equation will also be one that can always be characterized as of the [_____] degree.

_ _ _ _ _ _ _ _ _ _ _ _ _ _ _ _ _ _

first

A first-degree differential equation will also necessarily be a linear differential equation. [true, false]

_ _ _ _ _ _ _ _ _ _ _ _ _ _ _ _ _

false

Consider the equation

$$2\frac{d^2z}{dx^2} + \left(\frac{dz}{dx}\right)^2 - 4z = 0 \qquad (1\text{-}2)$$

This is an equation of the [_____] degree and is [linear, nonlinear].

_ _ _ _ _ _ _ _ _ _ _ _ _ _ _ _ _ _

first, nonlinear

Reference Equation

1-1 $\dfrac{d^2y}{dt^2} + 6\dfrac{dy}{dt} + 5y = 0$

5. Next, we need to distinguish between differential equations with constant coefficients and those with variable coefficients. The *coefficients* referred to in this terminology are the coefficients of the dependent variable and its derivatives, not the coefficients, if any, of terms in the *independent* variable. It should be easy, once the relevant coefficients are identified, to tell whether they are variables or constants. Try the following examples:

Equation (1-1) is an example of a linear differential equation with [constant, variable] coefficients.

— — — — — — — — — — — — — — — — — —

constant

Examine the equation

$$\dfrac{d^2x}{dt^2} + 52\left(\dfrac{dx}{dt}\right)^2 + 310x = t^2 \sin 377t$$

This equation is an example of a differential equation with [constant, variable] coefficients.

— — — — — — — — — — — — — — — — — —

constant

6. In our review of terminology, we also must treat the term "order." The order of a differential equation is [_____].

— — — — — — — — — — — — — — — — — —

the order of the highest derivative which appears in the equation after any integrals in the dependent variable have been removed by differentiation.

Equation (1-1) is of the [_____] order.

— — — — — — — — — — — — — — — — — —

second

What is the order of the following equation?

$20\dfrac{di}{dt} + 10i + 10 \int_0^t \dfrac{idt}{C} = 10$ (*C* is a constant)   [_____]

— — — — — — — — — — — — — — — — — —

second

Reference Equation

1-1 $\dfrac{d^2y}{dt^2} + 6\dfrac{dy}{dt} + 5y = 0$

---

7. The equations that we have used as differential equation examples have all been "ordinary" differential equations. The term "ordinary" is used in contrast with the term "partial" differential equations. Since this text is concerned with the former class only, we need to distinguish between these classes. The distinction is based on the kind of derivatives, either ordinary or partial, that appear in the equation.

Ordinary differential equations contain only [_____] derivatives.

– – – – – – – – – – – – – – – – – –

ordinary

Partial differential equations contain [_____] derivatives.

– – – – – – – – – – – – – – – – – –

partial

Partial differential equations contain at least some partial derivatives.

Equation (1-1) is an example of a differential equation which is [ordinary, partial].

– – – – – – – – – – – – – – – – – –

ordinary

> **Reference Equation**
>
> **1-2** $2\dfrac{d^2z}{dx^2} + \left(\dfrac{dz}{dx}\right)^2 - 4z = 0$

8. Last, we need to define the term "homogeneous" as applied to an ordinary differential equation.

In the sense used in this text, a linear differential equation is *homogeneous* when at least one derivative of the dependent variable or the dependent variable itself appears in each term of the equation.

Is Equation (1-1) homogeneous? [____]

- - - - - - - - - - - - - - - - - -

yes

Is Equation (1-2) homogeneous? [____]

- - - - - - - - - - - - - - - - - -

yes

Is the following equation homogeneous?

$0.1\dfrac{d^2z}{dt^2} + 14\dfrac{dz}{dt} + 5 \times 10^4 z = 141 \sin 377t$ [____]

- - - - - - - - - - - - - - - - - -

no

Is the following equation homogeneous?

$\dfrac{d^4x}{dt^4} + \left(\dfrac{d^2x}{dt^2}\right)^2 + \dfrac{1}{t}\dfrac{dx}{dt} = 0$ [____]

- - - - - - - - - - - - - - - - - -

yes

EXPLANATION: There is another meaning of homogeneous. This meaning relates to first-order differential equations. A first-order differential equation is said to be homogeneous when the equation can be written in the form:

$$\frac{dy}{dx} = \frac{M(x,y)}{N(x,y)}$$

and the functions M and N are homogeneous of the same order in the two variables x and y; i.e.,

$$M(tx,ty) = t^n M(x,y)$$
$$N(tx,ty) = t^n N(x,y)$$

In this case, the equation can also be written dy/dx = f(y/x). Homogeneity of this class is not relevant to this text, however.

Exercises

Several equations follow. Characterize each of these regarding their order, whether they are linear or nonlinear, have variable or constant coefficients, are ordinary or partial, are homogeneous or inhomogeneous, and their degree:

E–1 $\dfrac{d^2y}{dx^2} + 10y = 0$

[_____] order [linear, nonlinear] [variable coefficients, constant coefficients] [ordinary, partial] [homogeneous, inhomogeneous]
[_____] degree

－－－－－－－－－－－－－－－－－－

This is second order, linear, has constant coefficients, is ordinary, homogeneous, and of the first degree.

E–2 $\dfrac{d^4y}{dt^4} + 4\dfrac{d^3y}{dt^3} + 0.1\dfrac{d^2y}{dt^2} + 3\dfrac{dy}{dt} + 14y = 5 \sin t$

[_____] order [linear, nonlinear] [variable, constant] coefficients [ordinary, partial] [homogeneous, inhomogeneous] [_____] degree

－－－－－－－－－－－－－－－－－－

This is fourth order, linear, has constant coefficients, is ordinary, inhomogeneous, and of the first degree.

E–3 $\dfrac{d^2y}{dx^2} + \dfrac{1}{x}\dfrac{dy}{dx} + 14y = 10$

[_____] order [linear, nonlinear] [variable, constant] coefficients [ordinary, partial] [homogeneous, inhomogeneous] [_____] degree

－－－－－－－－－－－－－－－－－－

This is second order, linear, has variable coefficients, is ordinary, inhomogeneous, and of the first degree.

E–4 $\dfrac{d^3x}{dt^3} + 7\dfrac{d^2x}{dt^2} + 4\left(\dfrac{dx}{dt}\right)^2 + 5x = 1.02$

[_____] order [linear, nonlinear] [variable, constant] coefficients [ordinary, partial] [homogeneous, inhomogeneous] [_____] degree

－－－－－－－－－－－－－－－－－－

This is third order, nonlinear, has constant coefficients, is ordinary, inhomogeneous, and of the first degree.

E–5 $\quad \dfrac{\partial^2 e}{\partial x^2} = rge + (rc + gl)\dfrac{\partial e}{\partial t} + lc\dfrac{\partial^2 e}{\partial t^2}$

where r, g, l, c are constants.

[_____] order [linear, nonlinear] [variable, constant] coefficients
[ordinary, partial] [homogeneous, inhomogeneous] [_____] degree

- - - - - - - - - - - - - - - - - -

This is second order, linear, has constant coefficients, is partial,
homogeneous, and of the first degree.

E–6 $\quad 14\left(\dfrac{d^5 x}{dy^5}\right)^3 + 12\dfrac{d^4 x}{dy^4} + 6\dfrac{d^3 x}{dy^3} + 2.1\dfrac{d^2 x}{dy^2} = 0$

[_____] order [linear, nonlinear] [variable, constant] coefficients
[ordinary, partial] [homogeneous, inhomogeneous] [_____] degree

- - - - - - - - - - - - - - - - - -

This is fourth (fifth) order, is nonlinear, has constant coefficients,
ordinary, homogeneous, and of the third degree.

An explanation of the answer regarding the *order* given for this example follows:

The last equation differs from the other examples in that there is
no term in the dependent variable x itself. This "*apparently*" fifth-
order equation can be reduced to a *fourth*-order equation (in a different variable, to be sure) by the substitution:

$v = [$ _____ $]$

- - - - - - - - - - - - - - - - -

$v = \dfrac{dx}{dy}$

With this substitution, the equation becomes

$$14\left(\dfrac{d^4 v}{dy^4}\right)^3 + 12\dfrac{d^3 v}{dy^3} + 6\dfrac{d^2 v}{dy^2} + 2.1\dfrac{dv}{dy} + 5v = 0$$

After such an equation is solved for v, a further step will permit
solution for x. *Students are strongly urged to perform such a reduction of order, if possible, in any differential equation to be solved,
before proceeding with the rest of the solution.*

Summary of Section 1—Terminology

In Section 1, differential equations have been classified in the categories:

(1) order

(2) linear or nonlinear

(3) constant or variable coefficients

(4) ordinary or partial

(5) homogeneous or otherwise

(6) degree

This scheme of classification will enable the student to determine whether any particular differential equation lies within the basic categories to be studied in this work, so that the methods described herein can be applied.

Consider the solution of the very simplest differential equations, *ordinary* differential equations, which are *linear,* with *constant coefficients,* and *homogeneous.*

Section 2   AN EXAMPLE SOLUTION BY "CUT AND TRY" OF A SECOND-ORDER, LINEAR, HOMOGENEOUS, ORDINARY DIFFERENTIAL EQUATION WITH CONSTANT COEFFICIENTS

In this section, we will perform some "cut and try" operations on a simple differential equation to illustrate appropriate properties of a solution or solutions. After this specific example, we will give you a chance to try some generalizations.

1. The differential equation on which we will practice is Eq. (1-1), which was

$$\frac{d^2y}{dt^2} + 6\frac{dy}{dt} + 5y = 0 \tag{1-1}$$

As you will recall, this is a second-order, linear, homogeneous, ordinary differential equation with constant coefficients. Is $y = 0$, a solution of this equation? [____]

— — — — — — — — — — — — — — — — — —

yes

Clearly, $f(x) = 0$, where $f(x)$ and x are the dependent and independent variables, respectively, will be a solution of any homogeneous differential equation. However, this trivial solution is not the only possible solution, and, moreover, it is only occasionally a useful one.

::

Reference Equation

1-1 $\dfrac{d^2y}{dt^2} + 6\dfrac{dy}{dt} + 5y = 0$

::

2. We will now look at some more interesting possible solutions of Eq. (1-1). Is the function $y = e^{-5t}$ a solution of Eq. (1-1)? [____]

— — — — — — — — — — — — — — — — — —

yes

If your answer is "yes," see whether your argument coincides with that given below. If your answer was "no," the explanation will give you an opportunity to see why "no" was incorrect.

EXPLANATION: The function, $y = e^{-5t}$ is a solution of (1-1) because $dy/dt = -5e^{-5t}$, $d^2y/dt^2 = 25e^{-5t}$ and when y, dy/dt and d^2y/dt^2 are substituted in (1-1) we obtain [_____]

— — — — — — — — — — — — — — — — — —

$25e^{-5t} - 30e^{-5t} + 5e^{-5t} = 0$

Or, in other words, the result of substituting $y = e^{-5t}$ in Eq. (1-1) is that the equation is satisfied.

═══

3. Next, let us inquire whether the function $y = e^{-t}$ is a solution of the same Eq. (1-1). [yes, no]

— — — — — — — — — — — — — — — — —

yes

Is $y = 15e^{-t}$ also a solution of Eq. (1-1)? [____]

— — — — — — — — — — — — — — — — — —

yes

EXPLANATION: Skip this sentence if you are sure why "yes" is the correct response to the preceding frame. $y = 15e^{-t}$ is a solution because the result of substituting $y = 15e^{-t}$ in (1-1) gives [_____]

— — — — — — — — — — — — — — — — —

$15e^{-t} - 90e^{-t} + 75e^{-t} = 0$ or $0 \equiv 0$

═══

Reference Equation

1-1 $\dfrac{d^2y}{dt^2} + 6\dfrac{dy}{dt} + 5y = 0$

---

4. In fact, the function $y = ke^{-t}$, in which k has any constant value, is a solution of (1-1).

   Is this also true of $y = ke^{-5t}$? [＿＿]

— — — — — — — — — — — — — — — — — —

yes

---

5. Since the two functions, $y = ke^{-5t}$ and ke^{-t}, have both easily been shown to be valid solutions of Eq. (1-1), we might wonder whether there are any more solutions than these two solutions of exponential form. We check this by trying $y = ke^{mt}$ as a solution. Substitution of this function in Eq. (1-1) should give us an equation from which we can determine all the values of m for exponential functions, $y = ke^{mt}$, which solve Eq. (1-1). We already know of two values of m in the function $y = ke^{mt}$ that make this a solution of Eq. (1-1).

   When $y = ke^{mt}$ is substituted in Eq. (1-1) we obtain the equation

[＿＿＿＿＿＿＿＿＿＿＿＿＿＿＿＿＿＿＿＿] $= 0$

— — — — — — — — — — — — — — — — — —

$km^2e^{mt} + 6\,kme^{mt} + 5\,ke^{mt} = 0$ (1-3)

   There are several possible solutions of (1-3). Two of these are $k = 0$ and $e^{mt} = 0$. These are not meaningful solutions. If these trivial cases are eliminated, we are left with an equation in m, which is [＿＿＿＿＿＿＿＿＿＿＿＿] $= 0$

— — — — — — — — — — — — — — — — — —

$m^2 + 6\,m + 5 = 0$

   This describes the possible values of m, which will make $y = e^{mt}$ a solution of Eq. (1-1). These values are $m = [\rule{1cm}{0.4pt}\,,\,\rule{1cm}{0.4pt}]$.

— — — — — — — — — — — — — — — — — —

$m = -1,\ -5$

Thus, there are only two solutions of exponential form for Eq. (1-1); these are $y = ke^{-t}$ and $y = ke^{-5t}$. These we had previously found to be valid, although it was not apparent at the time that we tried these solutions why these were the only appropriate trial choices.

---

6. We have spent considerable time on these solutions of Eq. (1-1). One more important question should be answered. Is the function $y = k_1 e^{-t} + k_2 e^{-5t}$ also a solution of Eq. (1-1)? [____]

— — — — — — — — — — — — — — — — — — — —

yes

---

7. *Remarks.* The solution $y = k_1 e^{-t} + k_2 e^{-5t}$ is the most general solution for Eq. (1-1). The necessary values of the exponents of the two exponential terms are determined in this solution, but the values of the coefficients k_1 and k_2 are not. (Any values of k_1 and k_2 can be used.) It will develop that the values of the coefficients, i.e., *the constants k_1 and k_2, depend on the physical situation, for which Eq. (1-1) is but part of the mathematical model, and that these coefficients are as yet undetermined in this case because we have given an equation without indicating a physical situation to which it is relevant.* Incidentally, it can be shown that exponential functions are the only functions that satisfy this Eq. (1-1), i.e., we have not only found a most general exponential solution, but also the only possible form of the general solution to Eq. (1-1).

Section 3 A FIRST GENERALIZATION CONCERNING THE SOLUTION OF A SECOND-ORDER, LINEAR, ORDINARY DIFFERENTIAL EQUATION WITH CONSTANT COEFFICIENTS

1. We have been working with an equation of the form:

$$A\frac{d^2y}{dt^2} + B\frac{dy}{dt} + Cy = 0$$

On the basis of our work to this point, it would appear that the equation will have a solution in the form:

$$y = [\underline{\qquad\qquad}] + [\underline{\qquad\qquad}]$$

- - - - - - - - - - - - - - - - - - -

$$y = k_1 e^{m_1 t} + k_2 e^{m_2 t}$$

___

in which the values of the constants m_1 and m_2 can be determined by the following steps:

(a) Substitute into the equation the trial solution $y = [\underline{\qquad\qquad}]$

- - - - - - - - - - - - - - - - - - -

$$y = k\, e^{mt}$$

___

(b) Solve the resulting quadratic equation in "m" for m:

$$[\underline{\qquad\qquad\qquad}] = 0$$

- - - - - - - - - - - - - - - - - - -

$$Am^2 + Bm + C = 0$$

___

In general, this equation has two roots, m_1 and m_2. These are easily found and enable us to complete the solution $y = k_1 e^{m_1 t} + k_2 e^{m_2 t}$ as much as is possible at this stage. The coefficients k_1 and k_2 *can be determined only after knowing some further information not contained in the differential equation itself; this is the physical constraints (boundary conditions) of a problem,* only part of whose mathematical model is the differential equation itself.

___

2. We will find that *this first generalization* concerning the solution of the second-order, linear, ordinary differential equation with constant coefficients *is a good one, but not completely satisfactory.*

___

**Section 4 SOME PRACTICE WITH THE FIRST GENERAL-
IZATION APPLIED TO A FIRST-ORDER EQUATION**

1. Let us try our method on the equation

$$3\frac{dy}{dx} + 2y = 0 \tag{1-4}$$

What is your solution? [_____]

– – – – – – – – – – – – – – – – – –

$$y = ke^{-2/3x}$$

EXPLANATION: We substitute in Eq. (1-4) the function

$$y = [\text{_____}]$$

– – – – – – – – – – – – – – – – –

$$y = ke^{mx} \tag{1-5}$$

*Substitution of Eq. (1-5) in Eq. (1-4) yields the equation
$3m + 2 = 0$. The solution of this is $m = -2/3$. Thus we have as
solutions of Eq. (1-4): $y = 0$, $y = e^{-2/3x}$, $y = ke^{-2/3x}$, of which
the last is the most general.*

**Section 5 SOME FURTHER PRACTICE WITH THE FIRST
GENERALIZATION ON A HIGHER-ORDER EQUATION**

1. Find the solution of the following differential equation

$$\frac{d^4x}{dt^4} - 13\frac{d^2x}{dt^2} + 36x = 0 \tag{1-6}$$

Your best solution of this equation is [_____
_____]. The correct best solution is

– – – – – – – – – – – – – – – – – –

$$x = k_1e^{2t} + k_2e^{-2t} + k_3e^{3t} + k_4e^{-3t}$$

If you have obtained the foregoing solution (or anything equivalent
to it) skip the following explanation and turn to Frame 1 on page 17.
Otherwise, refer to the following frame.

Reference Equation

1-6 $\dfrac{d^4x}{dt^4} - 13\dfrac{d^2x}{dt^2} + 36x = 0$

EXPLANATION:

2.   *Steps for finding the solution of (1-6)*

Substitute for the dependent variable x in Eq. (1-6) the function
x = ke^{mt}. This gives the equation [_____
_____].

- - - - - - - - - - - - - - - - - - - -

$km^4 e^{mt} - 13km^2 e^{mt} + 36ke^{mt} = 0$

Since e$^{mt} \neq 0$, we can divide by e^{mt} to obtain the equation
$m^4 - 13m^2 + 36 = 0$ *or* $(m^2 - 9)$ [_____] $= 0$

- - - - - - - - - - - - - - - - - -

$(m^2 - 9)(m^2 - 4) = 0$ *or further,* $(m-3)(m+3)(m-2)(m+2) = 0$

---

so that m can be m = [_____].

- - - - - - - - - - - - - - - - - -

$m = +3, -3, +2, -2$

Thus, the four possible terms in our solution are
x = [_____] *or* [_____] *or* [_____] *or* [_____].

- - - - - - - - - - - - - - - - -

$x = k_1 e^{3t}$ *or* $k_2 e^{-3t}$ *or* $k_3 e^{2t}$ *or* $k_4 e^{-2t}$

The complete general solution is the sum of individual solutions
and, hence, is x = $k_1 e^{3t} + k_2 e^{-3t} + k_3 e^{2t} + k_4 e^{-2t}$. The con-
stant coefficients, k_1, k_2, k_3, and k_4, are as yet undetermined,
because we have not related our differential equation to any par-
ticular physical situation.

---

3. It is clear that so far our procedure for obtaining the exponential
solution works in every case in which we have tried it. In the next
section we will try to describe this procedure a little more generally
than we did in Section 3.

Section 6   FIRST TRIAL ATTEMPT AT A GENERALIZATION FOR THE SOLUTION OF THE nTH-ORDER, LINEAR, ORDINARY, HOMOGENEOUS DIFFERENTIAL EQUATION WITH CONSTANT COEFFICIENTS

1. Let us outline the steps which we have developed for solving a *linear, nth-order, homogeneous* differential equation with *constant coefficients.*

An equation of this type can be written as

$$a_n \frac{d^n y}{dx^n} + a_{n-1} \frac{d^{n-1} y}{dx^{n-1}} + \cdots + a_1 \frac{dy}{dx} + a_0 y = 0 \qquad (1\text{-}7)$$

Our first step in obtaining a solution is to substitute in this equation, the function $y = [\underline{\hspace{1cm}}]$.

- - - - - - - - - - - - - - - - - -

$y = e^{mx}$

Since $e^{mx} \neq 0$, this substitution gives the equation

$[\underline{\hspace{6cm}}] = 0$.

- - - - - - - - - - - - - - - - - -

$a_n m^n + a_{n-1} m^{n-1} + \cdots + a_1 m + a_0 = 0$

Incidentally, this polynomial equation, obtained by substituting $y = e^{mx}$ in the differential Equation (1-7), is called the *auxiliary equation.*

---

2. *Step II in the Solution:* Solve the auxiliary equation, the algebraic equation in "m". There can be at most "n" distinct roots which satisfy this auxiliary equation, in which n is the order of the equation. Incidentally, finding the roots of the auxiliary equation when this auxiliary equation is of a higher order than the second can prove quite difficult, especially when there are complex roots. Since this algebraic procedure forms a separate discipline by itself, we will not make a point of it at this time.

---

3. *Step III in the Solution:* If we term these "n" values of "m" as $m_1 \ldots m_n$, the solution of the homogeneous differential equation is

$[\underline{\hspace{6cm}}]$.

- - - - - - - - - - - - - - - - - -

$y = k_1 e^{m_1 x} + k_2 e^{m_2 x} + \cdots + k_n e^{m_n x} \qquad (1\text{-}8)$

---

4. The three preceding steps, Steps I, II, and III, would appear to constitute a general method of solution (at least, a solution except for finding the values of the coefficients k_1, k_2 ... k_n, a process which has not yet been discussed). We will see in Section 7, however, that this "general" method does not cover every possible contingency.

The contingency not covered is that in which the group of roots of the auxiliary equation includes some repeated roots.

Section 7   REPEATED ROOTS: A CASE NOT COVERED BY FIRST-TRIAL GENERALIZATION AND A NEW GENERALIZATION

1. Under certain circumstances, some of the values of "m" may be equal to one another. For example, in the differential equation

$$\frac{d^2y}{dx^2} + 2\frac{dy}{dx} + y = 0 \tag{1-9}$$

the corresponding auxiliary equation is [_____]

- - - - - - - - - - - - - - - - - -

$m^2 + 2m + 1 = 0$

---

and the roots are $m = $ [_____, _____]

- - - - - - - - - - - - - - - - - -

$m = -1, -1$

According to our preceding "first-order" generalization, the solution of Eq. (1-9) would be $y = $ [_____ _____].

- - - - - - - - - - - - - - - - - -

$y = k_1 e^{-x} + k_2 e^{-x} = e^{-x}(k_1 + k_2) = k_3 e^{-x}$

---

2. Instead of two undetermined constant coefficients of exponential terms in the solution, we actually have in effect but one. For reasons that will become apparent later, the general solution of a second-order linear differential equation with constant coefficients must have *two* undetermined constants. The general solution of an *n*th-order linear differential equation with constant coefficients must have [____] undetermined coefficients.

- - - - - - - - - - - - - - - - - -

n

Thus we have to seek some other form of general solution for the case of the homogeneous equation which has repeated roots.

Reference Equation

1-9 $\dfrac{d^2y}{dx^2} + 2\dfrac{dy}{dx} + y = 0$

---

3. We will proceed to demonstrate the proper solution of this equation by the same procedure used with earlier cases which did not have repeated roots. (The procedure will be the same but the solution will be different!)

Is $y = k_4 e^{-x}$ a solution of Eq. (1-9)? [____]

— — — — — — — — — — — — — — — — — —

yes

This function, $y = k_4 e^{-x}$, has already been shown to be a solution of Eq. (1-9). However, it has been stated that we need *another* solution in addition to this function.

Is $y = k_2\, x e^{-x}$ a solution of Eq. (1-9)? [____]

— — — — — — — — — — — — — — — — — —

yes

Is the function

$$y = k_1 e^{-x} + k_2 x e^{-x} \qquad (1\text{-}10)$$

a solution of Eq. (1-9)? [____]

— — — — — — — — — — — — — — — — — —

yes

Thus we have in Eq. (1-10) a solution with the required *two* undetermined, independent coefficients. Although we have not proved it, it is also true that Eq. (1-10) is the complete general solution; there are no other possible terms.

---

4. Before trying a generalization for the case of any auxiliary equation with repeated roots, let us look at another case.

Consider the differential equation

$$\frac{d^3y}{dt^3} + 4\frac{d^2y}{dt^2} + 5\frac{dy}{dt} + 2y = 0 \qquad (1\text{-}11)$$

The roots of the auxiliary equation are $m = -1, -1, -2$.
Is $y = k_1e^{-t} + k_2te^{-t} + k_3e^{-2t}$ a solution of this equation?
[_____]

- - - - - - - - - - - - - - - - - -

yes

_____

Is $y = k_1e^{-t} + k_2te^{-2t} + k_3te^{-2t}$ a solution of Eq. (1-11)?
[_____]

- - - - - - - - - - - - - - - - - -

no

_____

5. Evidently, when there are more than two roots of the auxiliary equation $m = d_1, d_2, d_3, d_4, \ldots d_n$, and there is at least one pair of repeated roots, such as $d_3 = d_4$, the solution of the differential equation will have the form

$$y = k_1e^{d_1t} + k_2e^{d_2t} + k_3e^{d_3t} + k_4te^{d_4t} + \cdots k_ne^{d_nt}$$

For every pair of repeated roots, $m = d_i, d_i$, there will be a pair of terms, one of which contains $y = k_ie^{d_it}$, just as in the case in which the root is not repeated, and the other of which repeats the exponential term with the independent variable in the coefficient, as in $y = [_____] e^{d_it}$.

- - - - - - - - - - - - - - - - - -

$k_ite^{d_it}$

6. Suppose that a root is repeated more than twice, as would be the case for the differential equation

$$\frac{d^4x}{dt^4} + 9\frac{d^3x}{dt^3} + 30\frac{d^2x}{dt^2} + 44\frac{dx}{dt} + 24x = 0 \qquad (1\text{-}12)$$

The auxiliary equation is [_____]$=0$

- - - - - - - - - - - - - - - - - - -

$$m^4 + 9m^3 + 30m^2 + 44m + 24 = 0$$

The roots of this equation are $m = -2, -2, -2, -3$. The solution is $x = $ [_____].

- - - - - - - - - - - - - - - - - - -

$$x = k_1 e^{-2t} + k_2 t e^{-2t} + k_3 t^2 e^{-2t} + k_4 e^{-3t}$$

The validity of this solution may be checked by substitution in the original equation, Eq. (1-12).

7. We are now ready to make a generalization for the solution of homogeneous, linear, ordinary differential equations with constant coefficients, regardless of whether there are repeated roots or not. We have already given the solution for the case without repeated roots (page 17). To this, we add the statement: When the auxiliary equation contains roots m_1, m_2, ... m_n, of which the roots m_1 through m_i are repeated with a value m_i, m_h, m_g ... $m_2 = m_1$, the solution will be

$$f(t) = k_1 e^{m_1 t} + [_____] + [_____] + \cdots + k_i t^{i-1} e^{m_i t}$$
$$+ k_{i+1} e^{m_i t} + \cdots [_____]$$

- - - - - - - - - - - - - - - - - - -

$$f(t) = k_1 e^{m_1 t} + k_2 t e^{m_1 t} + k_3 t^2 e^{m_1 t} + \cdots k_n e^{m_n t}$$

8. We may ask ourselves: "As a practical matter, just how important is the case of repeated roots?" This case is not very common. In practice, we find that equations which result in complex values of "m" are more common in engineering practice than those with real roots, a situation in which repeated roots are infrequent. Consequently, a repeated root does not occur very often in engineering problems. However, the case of repeated roots is not an unimportant one and it was necessary to consider it.

Section 8   PRACTICE WITH A PROBLEM TO WHICH THE REVISED GENERALIZATION IS APPLICABLE

1. For practice, try solving the following equation:

$$\frac{d^4x}{dt^4} - 8\frac{d^2x}{dt^2} + 16x = 0 \qquad (1\text{-}13)$$

The auxiliary equation is [_____]

- - - - - - - - - - - - - - - - - -

$$m^4 - 8m^2 + 16 = 0$$

which has the roots $m = [\_\_\_, \_\_\_, \_\_\_, \_\_\_]$.

- - - - - - - - - - - - - - - - - -

$$m = +2, -2, +2, -2$$

So, the foregoing equation has the solution $x = [$_____
_____]

- - - - - - - - - - - - - - - - - -

$$x = k_1e^{2t} + k_2te^{2t} + k_3e^{-2t} + k_4te^{-2t}$$

Section 9   SOME REFINEMENTS IN SOLUTIONS OF HOMO-GENEOUS DIFFERENTIAL EQUATIONS IN WHICH THE AUXILIARY EQUATION HAS IMAGINARY OR COMPLEX ROOTS

Although the form of the solutions given up to this point is entirely accurate, it may not be the most convenient algebraic form for cases in which the auxiliary equation has imaginary or complex roots.

1. Let us consider the homogeneous differential equation

$$\frac{d^2z}{dx^2} + 4z = 0 \qquad (1\text{-}14)$$

This has the auxiliary equation [_____].

- - - - - - - - - - - - - - - - - -

$$m^2 + 4 = 0$$

The roots of the auxiliary equations are $m = [\_\_], [\_\_]$

- - - - - - - - - - - - - - - - - -

$$m = 2i, m = -2i$$

where $i = \sqrt{-1}$; since $(+2i)^2 = -4$ and $(-2i)^2 = -4$.

The general solution of the equation is, then, $z = [$_____$] + [$_____$]$

- - - - - - - - - - - - - - - - - -

$$z = k_1e^{+2ix} + k_2e^{-2ix} \qquad (1\text{-}15)$$

Reference Equation

1-15 $z = k_1 e^{+2ix} + k_2 e^{-2ix}$

---

2. The solution, Eq. (1-15), is algebraically correct. However, we usually prefer not to work with imaginary exponents and use, instead, a different form of solution based upon the identities:

$$e^{imx} = \cos mx + i \sin mx \qquad\qquad (1\text{-}16)$$

$$e^{-imx} = [\underline{\quad\quad} + \underline{\quad\quad}] \qquad\qquad (1\text{-}17)$$

- - - - - - - - - - - - - - - - - -

$e^{imx} = \cos mx - i \sin mx$

---

3. Using the identities (1-16) and (1-17), the solution can alternatively be written as $z = [\underline{\qquad\qquad\qquad\qquad\qquad\qquad\qquad}]$.

- - - - - - - - - - - - - - - - - -

$$z = (k_1 + k_2) \cos 2x + i (k_1 - k_2) \sin 2x \qquad\qquad (1\text{-}18)$$

Moreover, we can combine the coefficients in Eq. (1-18) to obtain $k_1 + k_2 = k_3$, and $i(k_1 - k_2) = k_4$.

Without loss of generality, because only two undetermined coefficients are required for a second-order equation, the solution can then be written as $z = [\underline{\qquad\quad}] + [\underline{\qquad\quad}]$.

- - - - - - - - - - - - - - - - - -

$$z = k_3 \cos 2x + k_4 \sin 2x \qquad\qquad (1\text{-}19)$$

Remark. The validity of the trigonometric form (1-19) of the solution can readily be verified by substituting it directly in the original differential equation.

---

4. The foregoing example was a case with imaginary roots. Let us look at a differential equation for which the auxiliary equation has complex roots:

$$\frac{d^2x}{dt^2} + 2\frac{dx}{dt} + 2x = 0 \qquad (1\text{-}20)$$

The auxiliary equation is [_____].

- - - - - - - - - - - - - - - - -

$m^2 + 2m + 2 = 0$

_____

The roots are $m = [_____]$; $m = [_____]$.

- - - - - - - - - - - - - - - - -

$m = -1 + i,\ \ m = -1 - i$

_____

Thus the solution could be written as $x = [_____$
$_____]$.

- - - - - - - - - - - - - - - - -

$$x = k_1 e^{(-1+i)t} + k_2 e^{(-1-i)t} = e^{-t}(k_1 e^{it} + k_2 e^{-it}) \qquad (1\text{-}21)$$

---

5. However, Eq. (1-21) can be written more conveniently in terms of a combination of a coefficient with a real exponent multiplied by the sum of terms in sines and cosines.
In that form it is $x = [_____]$.

- - - - - - - - - - - - - - - - -

$x = e^{-t}(k_3 \cos t + k_4 \sin t)$

If your answer on the above line was correct, skip the following section and go directly to Frame 6.

> *EXPLANATION: The equation (1-21) can be factored to obtain*
> $x = e^{-t}(k_1 + k_2) \cos t + i(k_1 - k_2) \sin t$ *and, by substituting*
> $k_1 + k_2 = k_3$ *and* $i(k_1 - k_2) = k_4$, *we have* $x = e^{-t}(k_3 \cos t + k_4 \sin t)$.

Remark. The validity of this solution can readily be verified by substituting it in the original equation, Eq. (1-20).

6. For practice with finding the convenient trigonometric forms when there are complex roots, let us look at the solution of the following differential equation:

$$\frac{d^4y}{dx^4} - 7\frac{d^3y}{dx^3} + 18\frac{d^2y}{dx^2} - 22\frac{dy}{dx} + 12y = 0 \qquad (1\text{-}22)$$

This is a fourth-order homogeneous linear differential equation, with constant coefficients. The auxiliary equation is

[_____].

- - - - - - - - - - - - - - - - - -

$$m^4 - 7m^3 + 18m^2 - 22m + 12 = 0 \qquad (1\text{-}23)$$

This auxiliary equation will have [four, less than four] roots.

- - - - - - - - - - - - - - - - - -

four

Reference Equations

1-22　$\dfrac{d^4y}{dx^4} - 7\dfrac{d^3y}{dx^3} + 18\dfrac{d^2y}{dx^2} - 22\dfrac{dy}{dx} + 12y = 0$

1-23　$m^4 - 7m^3 + 18m^2 - 22m + 12 = 0$

7. Since the procedure for finding the roots of such a fourth-order equation is quite tedious, you are advised that two of the four roots are $m = 2$, $m = 3$. The remaining two roots, then, can be found quite easily.

There are $m = [\underline{\hspace{2cm}}]$ and $m = [\underline{\hspace{2cm}}]$.

- - - - - - - - - - - - - - - - - -

$m = 1 + i$　and　$m = 1 - i$

___

If your results are incorrect or you do not know how to obtain this result, see the following explanation. Otherwise skip to Frame 8.

EXPLANATION: Let us call the unknown roots m_1 and m_2; the auxiliary equation was

$$m^4 - 7m^3 + 18m^2 - 22m + 12 = 0 \qquad (1\text{-}23)$$

Now the left-hand side of this equation can be factored as

$[\underline{\hspace{1cm}}\ \underline{\hspace{1cm}}\ \underline{\hspace{1cm}}\ \underline{\hspace{1cm}}] = m^4 - 7m^3 + 18m^2 - 22m + 12$

- - - - - - - - - - - - - - - - - -

$(m - m_1)(m - m_2)(m - 2)(m - 3) = m^4 - 7m^3 + 18m^2 - 22m + 12$

___

If we divide both sides of the equation by the product $(m - 2)$ $(m - 3)$, we obtain

$$(m - m_1)(m - m_2) = \frac{m^4 - 7m^3 + 18m^2 - 22m + 12}{(m - 2)\,(m - 3)}$$

Since the denominator compromises only expressions in roots of numerator, there can be no remainder after the division and the quotient is $[\underline{\hspace{2cm}}]$.

- - - - - - - - - - - - - - - - - -

$m^2 - 2m + 2$

___

This is only a second-order equation. Using the quadratic formula, we see that the roots of this equation are

$m = [\underline{\hspace{2cm}}]$, *i.e.*, $m = [\underline{\hspace{2cm}}]$.

- - - - - - - - - - - - - - - - - -

$m = (2 \pm \sqrt{4 - 8}\,)/2$, *i.e.*, $m = 1 + i,\ 1 - i$

___

8. Therefore the solution of the fourth-order differential equation (1-22) can be written as

$$y = [\underline{\qquad}] + [\underline{\qquad}] + [\underline{\qquad}] + [\underline{\qquad}].$$

$$y = k_1 e^{(1+i)x} + k_2 e^{(1-i)x} + k_3 e^{2x} + k_4 e^{3x}$$

　　The complex exponential terms can be more conveniently written in the trigonometric form so as to obtain $y = [\underline{\qquad}$
$\underline{\qquad}].$

$$y = e^x (k_5 \cos x + k_6 \sin x) + k_3 e^{2x} + k_4 e^{3x}$$

Remarks. Many differential equations arising in practical problems involve complex roots only and few, if any, ever have only simple integral real roots. It should be remembered that the *complex roots always occur in conjugate pairs.*

---

9. Try this equation for further practice:

$$\frac{d^4x}{dt^4} + 2\frac{d^3x}{dt^3} - 2\frac{d^2x}{dt^2} + 8x = 0 \qquad (1\text{-}24)$$

The corresponding auxiliary equation is $[\underline{\qquad}] = 0.$

$$m^4 + 2m^3 - 2m^2 + 8 = 0 \qquad (1\text{-}25)$$

The auxiliary equation (1-25) has four roots, one of which is $m = 1 + i$. Then, another root must be $[\underline{\qquad}].$

$$m = 1 - i$$

The others are, then, $m = [\underline{\qquad}]$ and $m = [\underline{\qquad}].$

$m = -2$　and　$m = -2$ (both real)

The solution, written in trigonometric terms, is
$$x = [\underline{\qquad\qquad}].$$

$$x = e^t (k_1 \cos t + k_2 \sin t) + e^{-2t} (k_3 + k_4 t)$$

Remark. The second bracket contains a term in $k_4 t$, since the root $m = -2$ is repeated.

10. Finally, as a last bit of practice, obtain the general solution for the equation

$$\frac{d^4y}{dt^4} - 8\frac{d^3y}{dt^3} + 42\frac{d^2y}{dt^2} - 104\frac{dy}{dt} + 169y = 0 \qquad (1\text{-}26)$$

Hint: Here, one of the four roots is $m_1 = 2 - i3$. Then, another root must be $m_2 = [\underline{\hspace{2cm}}]$.

- - - - - - - - - - - - - - - - - - - -

$m_2 = 2 + i3$

- - - - - - - - - - - - - - - - - - - -

Then the remaining two roots are $m_3 = [\underline{\hspace{1.5cm}}]$ and $m_4 = [\underline{\hspace{1.5cm}}]$.

- - - - - - - - - - - - - - - - - - - -

$m = 2 + i3$ and $m = 2 - i3$

- - - - - - - - - - - - - - - - - - - -

Remark. Thus we see that the conjugate roots $2 \pm i3$ are each repeated twice. The exponential solution is

$y = [\underline{\hspace{1.5cm}}] + [\underline{\hspace{1.5cm}}] + [\underline{\hspace{1.5cm}}] + [\underline{\hspace{1.5cm}}]$.

- - - - - - - - - - - - - - - - - - - -

$y = k_1 e^{(2+i3)t} + k_2 te^{(2+i3)t} + k_3 e^{(2-i3)t} + k_4 te^{(2-i3)t}$

- - - - - - - - - - - - - - - - - - - -

Where $k_5 = k_1 + k_3$ $k_6 = k_2 + k_4$
$\qquad\quad k_7 = i(k_1 - k_3)$ $k_8 = i(k_2 - k_4)$,
the general solution of Eq. (1-26) in trigonometric form is,

$y = [\underline{\hspace{3cm}}] + [\underline{\hspace{3cm}}]$.

- - - - - - - - - - - - - - - - - - - -

$y = (k_5 + k_6 t)e^{2t}\cos 3t + (k_7 + k_8 t)e^{2t}\sin 3t$

Section 10   SUMMARY OF METHOD OF SOLUTION OF HOMOGENEOUS OLDECC TO THIS POINT

We will again summarize the steps that we have developed for solving an nth-order OLDECC. An equation of this type can be written as

$$a_n\frac{d^n y}{dx^n} + a_{n-1}\frac{d^{n-1}y}{dx^{n-1}} : \ldots a_1\frac{dy}{dx} + a_0 y = 0$$

The first step in the solution is to

$[\underline{\hspace{7cm}}]$.

- - - - - - - - - - - - - - - - - - - -

substitute in the equation the function $y = e^{mx}$

(continued opposite page)

This gives us an equation called the [_____].

- - - - - - - - - - - - - - - - - -

auxiliary equation

The auxiliary equation for the foregoing OLDECC is

[_____].

- - - - - - - - - - - - - - - - -

$a_n m^n + a_{n-1} m^{n-1} \ldots a_1 m + a_0 = 0$

The next step is to [_____].

- - - - - - - - - - - - - - - - -

solve for the roots of the auxiliary equation

The auxiliary equation can have, at most, n distinct roots. There are several possibilities as to the number and kind of roots. For instance, some of the roots may be repeated. Moreover, the roots can be real, imaginary, or complex. The following exercises may be useful in summarizing our conclusions concerning the forms of the general solution for the various forms and numbers of roots.

In the following table, indicate which of the terms in the right-hand column are appropriate for the numbered classes in the left-hand column, i.e., for each item in the left-hand column, select the appropriate form of solution from the right-hand column. Write the corresponding number in the blank in the bracket on the left. All correct answers are at the bottom below the double line. Fill in all brackets before checking any answers.

TYPE OF ROOTS OF THE AUXILIARY EQUATION	*TYPICAL TERMS IN THE GENERAL SOLUTION*
[____] Real roots, single (i.e., *not* repeated)	(1) $k_1 \cos at + k_2 \sin at$
[____] Real roots, repeated	(2) $(k_1 \cos at + k_2 \sin at)e^{-bt}$
[____] Imaginary roots, single	(3) $(k_1 + k_2 t)e^{-at}$
[____] Complex roots, single	(4) $k_1 e^{-at} + k_2 e^{-bt} + k_3 e^{-ct}$
[____] Complex roots, repeated	(5) $e^{-at}[k_1 \cos bt + k_2 \sin bt] + te^{-at}[k_3 \cos bt + k_4 \sin bt]$

- - - - - - - - - - - - - - - - - - -

(4)
(3)
(1)
(2)
(5)

Now it is appropriate to go into the basis for determination of the so far undetermined coefficients of the exponential or trigonometric terms in the solutions. This procedure is carried out in Part B.

Part B—Procedure for Determination of Coefficients of Terms in the Solution of Homogeneous Differential Equations

Section 1   GENERAL PROCEDURE

As has been stated earlier, although the solutions up to this point have been valid for any value of the coefficients, in practice the coefficients actually have definite values that depend upon constraints in the physical situation from which the differential equations have been derived. These definite coefficient values depend upon *"initial conditions."*

---

1. For instance, in the solution $y = k_1 e^{-t} + k_2 e^{-5t}$ if we should somehow find out that when

$$t = 0, \quad y = 10 \quad \text{and} \quad \frac{dy}{dt} = 2 \qquad (1\text{-}27)$$

we would know that $k_1 = [\underline{\hspace{1cm}}]$, $k_2 = [\underline{\hspace{1cm}}]$.

- - - - - - - - - - - - - - - - - - - -

$k_1 = 13$ and $k_2 = -3$

Skip the following explanation if the values of k_1 and k_2 that you selected are correct.

> *EXPLANATION: These results are obtained as follows: for t = 0, the values of y and dy/dt from the foregoing equation are y = [\underline{\hspace{2cm}}] and dy/dt = [\underline{\hspace{2cm}}]*

- - - - - - - - - - - - - - - - - - - -

$y = k_1 + k_2$ and $\dfrac{dy}{dt} + -k_1 - 5k_2$

and the simultaneous solution of these two equations for y = 10 and dy/dt = 2 yields $k_1 = 13$, $k_2 = -3$.

---

2. The pair of equations (1-27) is a set of what are termed *initial conditions.* Such conditions, if a sufficient number of them are known, always permit the determination of the otherwise undetermined coefficients in the solution of the homogeneous differential equation.

---

3. For a second-order equation:

one initial condition is enough [yes, no]

– – – – – – – – – – – – – – – – – – –

no

two initial conditions are always sufficient [yes, no]

– – – – – – – – – – – – – – – – – –

yes

or sometimes three initial conditions may be necessary [yes, no].

– – – – – – – – – – – – – – – – – –

no

For an nth-order equation, the number of initial condition equations required is [＿].

– – – – – – – – – – – – – – – – – –

n

Remarks. It should be made quite clear that determination of the initial conditions is impossible from the differential equation itself. One must refer to the physical situation for which the equation is developed. Simple examples of physical configurations, development of differential equation models, and initial conditions only will be given here; more extensive treatments of the subject appropriate to any field (e.g., electric circuits, mechanics, etc.) can be found in books specializing in that field.

Section 2   A SIMPLE EXAMPLE OF A SOLUTION OF A PHYSICAL PROBLEM WHICH IS SOLVED WITH A HOMOGENEOUS, ORDINARY, LINEAR DIFFERENTIAL EQUATION WITH CONSTANT COEFFICIENTS

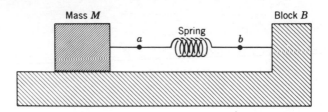

FIGURE 1-1 Sketch of a simple mass-spring system. Mass *M* is free to slide on a well-greased rigid horizontal surface. The *a* and *b* ends of the self-supporting spring are attached to mass *M* and a rigid block *B* respectively. If mass *M* is pulled away from its equilibrium position and then released. the mass *M* will slide back towards the equilibrium position under the influence of the force of the spring. It may well oscillate back and forth about the equilibrium position.

Let us consider the simple configuration shown in Fig. 1-1.

Let us take as the problem to be solved the analytical description of the motion of the block in this simple physical configuration after the mass *M* has been pulled out of its equilibrium position and then released.

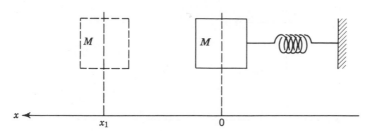

FIGURE 1-2 Suggested coordinates for the simple spring-mass problem. Displacement and forces to the left will be regarded as displacements and forces in the plus *x* direction. The equilibrium position (spring neither pulling nor pushing) will be taken as *x* = 0.

1. First, to describe the situation analytically, we need some coordinates, as shown in Fig. 1-2. These are a matter of choice, of
course.

   There are a number of possibilities as to the mathematical description of the force of the spring on the mass. We will assume as a
part of our problem statement that the spring has such properties
that its force is described by Hooke's Law.

   From Hooke's Law we know that the force on the end of the spring
and the displacement *x* of its end are [_____ _____].

----

directly proportional

---

2. The constant of proportionality will be termed k_s, so that the force
on the mass *M* due to the spring, will be:

   Force on mass = $[+, -] \, k_s x$

----

negative sign

   *EXPLANATION. The sign should be negative because the coor-
   dinates and directions were chosen in Fig. 1-2 in such a way
   that force is measured as positive when it acts to the left. A
   positive x displacement will result in a force that is actually acting to the right, or a force negative in sign for positive x.*

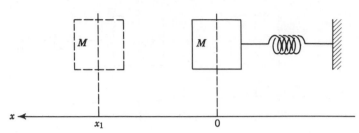

FIGURE 1-2 Suggested coordinates for the simple spring-mass problem. Displacement and forces to the left will be regarded as displacements and forces in the plus x direction. The equilibrium position (spring neither pulling nor pushing) will be taken as $x = 0$.

3. Let us next assume, for an initial attack, at least, that the *frictional forces* (mass M sliding on the surface, internal friction in the spring, wind resistance to motion, etc.) *are negligible.*

   Then there remains only one additional force in the system (in addition to the spring force); this is the force caused by the [acceleration, velocity, displacement, weight] of the mass.

   _____

   acceleration

   _____

   This force is

   $$F_M = [+, \ -] \ M \frac{d^2x}{dt^2}$$

   _____

   negative sign

   _____

   This force is negative because *positive accelerations to the left will produce forces to the right, thus the force is in a negative direction.*

4. Newton's Laws of Motion (in the form known as D'Alembert's Principle) tell us that the sum of the forces acting on the mass M must be zero, so that we obtain the differential equation

   $$[\underline{\hspace{4cm}}] = 0$$

   _____

   $$- M \frac{d^2x}{dt^2} - k_s x = 0 \tag{1-28}$$

   or, more conveniently, by multiplying by (-1), Eq. (1-28) is

   $$M \frac{d^2x}{dt^2} + k_s x = 0 \tag{1-29}$$

Reference Equation

1-29 $M \dfrac{d^2x}{dt^2} + k_s x = 0$

---

5. Eq. (1-29) is a homogeneous, ordinary, second-order, linear differential equation with constant coefficients. The auxiliary equation is [_____] = 0.

- - - - - - - - - - - - - - - - - - -

$Mm^2 + k_s = 0$ (1-30)

The roots of the auxiliary equation (1-30) are $m = $ [_____],
$m = $ [_____].

- - - - - - - - - - - - - - - - - - -

$m = \pm \sqrt{\dfrac{-k_s}{m}}$ or $m = i\sqrt{\dfrac{k_s}{m}}$ $m = -i\sqrt{\dfrac{k_s}{m}}$

The solution of the differential equation (1-29) in trigonometric form is $x = k_1 \cos$ [_____] + [_____].

- - - - - - - - - - - - - - - - - - -

$x = k_1 \cos \sqrt{\dfrac{k_s}{M}} t + k_2 \sin \sqrt{\dfrac{k_a}{M}} t$ (1-31)

---

6. It is part of our statement of the problem that initially mass M is pulled out to a position x_1. If we let the mass M be released at a time $t = 0$ (reference time), we have the initial condition that at $t = 0$, $x = $ [_____].

- - - - - - - - - - - - - - - - - - -

$x = x_1$ (1-32)

This is one initial condition for the physical system. One initial condition is enough to determine k_1 and k_2 in Eq. (1-31) [yes, no]?

- - - - - - - - - - - - - - - - - - -

no

Two initial conditions are required [yes, no].

- - - - - - - - - - - - - - - - - - -

yes

Occasionally, three initial conditions may be necessary [yes, no].

- - - - - - - - - - - - - - - - - - -

no (not for a 2nd-order equation)

---

7. A second initial condition would logically deal with the value, at $t = 0$, of [_____].

- - - - - - - - - - - - - - - - - - - -

$\dfrac{dx}{dt}$, velocity

What is this velocity? Actually the value of $\dfrac{dx}{dt}$ at $t = 0$ depends upon whether we simply release the mass at $t = 0$ (Case a) or give it on initial push at $t = 0$ (Case b) so that it moves with an initial velocity V_0.

In the first case (Case a), when $t = 0, \dfrac{dx}{dt} = [\underline{\quad\quad}]$ .... (1-33)

- - - - - - - - - - - - - - - - - - - -

zero

In the second case (Case b) $\dfrac{dx}{dt} = [\underline{\quad\quad}]$ at $t = 0$ .... (1-34)

- - - - - - - - - - - - - - - - - - - -

V_0

The first case, of course, corresponds to our initial statement of the problem, given on page 32.

Reference Equations

1-31 $x = k_1 \cos \sqrt{\dfrac{k_s}{M}}\, t + k_2 \sin \sqrt{\dfrac{k_s}{M}}\, t$

1-32 $x = x_1$

1-33 $\dfrac{dx}{dt} = 0$

---

8. We find from the initial conditions given by Eq. (1-32) and (1-33) that $k_1 = [\_\_\_\_\_]$ $k_2 = [\_\_\_\_\_]$.

$k_1 = x_1$ and $k_2 = $ zero

----

and our solution for the motion of the mass M in the problem is $x = [_____]$.

$x = x_1 \cos \sqrt{\dfrac{k_s}{M}}\, t$ (1-35)

According to this solution, the mass oscillates back and forth, indefinitely, sinusoidally in time, about its equilibrium position with a maximum displacement of $[\_\_\_\_]$.

----

x_1

----

and a frequency of oscillation f of $[_____]$.

$f = \dfrac{1}{2\pi} \sqrt{\dfrac{k_s}{M}}$ (1-36)

(Of course, we know that friction will actually cause the amplitude of the oscillation to gradually decay; however, the frequency calculated above might be substantially correct if the actual friction on the sliding mass is sufficiently small.)

---

9.   Other initial conditions are possible. For instance, as already suggested, we could give the mass an initial velocity V_0, starting at $x = x_1$. Or we could give the mass an initial velocity V_1, starting at $x = 0$. In each case the solution will be different because the initial conditions are different. However, the general solution (1-31) is correct in every case and must simply be completed with a choice of appropriate values of the coefficients k_1 and k_2.

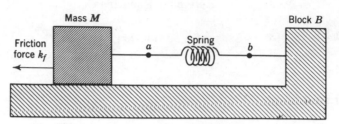

FIGURE 1-3 (figure 1-1 with friction force indicated).

10. Summary—We can see that the mathematical model of a physical problem of the type first described will include a [_____] equation;

————————————————

differential

————————————————

and independent of this, a statement of [_____ _____].

————————————————

initial condition(s).

————————————————

Both parts of the mathematical model are obtained by application of physical principles *and a careful examination of the problem statement.*

Section 3   A MORE INVOLVED EXAMPLE OF A PHYSICAL CONFIGURATION

1. Let us consider again the physical situation of Fig. 1-1 but this time attempt to take into account some frictional force (Fig. 1-3). Let us suppose that the most important "frictional" forces are such as to be reasonably taken into account by postulating a frictional force that is proportional to the velocity of mass M, i.e., that

$$\text{friction force} = [+, -]\, k_f \frac{dx}{dt} \tag{1-37}$$

Since the force will be to the right (negative direction) when dx/dt is positive (frictional force "holds" the mass back); the sign in Eq. (1-37) is negative.

2. The differential equation in this more involved situation is, then
[_____].

- - - - - - - - - - - - - - - - - -

$$- M \frac{d^2x}{dt^2} - k_f \frac{dx}{dt} - k_s x = 0 \qquad (1\text{-}38)$$

This is a more complicated differential equation than the differential equation in the preceding example. Is it still of the homogeneous type we have been considering? [yes, no]

- - - - - - - - - - - - - - - - - -

yes

To make the properties of this configuration more apparent, some actual values of the physical parameters will be given as a further part of the problem statement, as follows:

$M = 1$ kg;

$k_s = 6.25$ newtons/meter

$k_f =$ any one of the following three values, considered as three separate cases: 3,5, or 5.831 newtons/meter per second.

Three different values have been selected for k_f so that the student can see the effect of this friction parameter in three important and distinct cases. We thus have three different equations, obtained by substituting numerical values in Eq. (1-38)

Case A: $\quad \dfrac{d^2x}{dt^2} + 3\dfrac{dx}{dt} + 6.25x = 0 \qquad (1\text{-}39)$

Case B: $\quad \dfrac{d^2x}{dt^2} + 5\dfrac{dx}{dt} + 6.25x = 0 \qquad (1\text{-}40)$

Case C: $\quad \dfrac{d^2x}{dt^2} + 5.831\dfrac{dx}{dt} + 6.25x = 0 \qquad (1\text{-}41)$

Reference Equations

1-39 $\dfrac{d^2x}{dt^2} + 3\dfrac{dx}{dt} + 6.25x = 0$

1-40 $\dfrac{d^2x}{dt^2} + 5\dfrac{dx}{dt} + 6.25x = 0$

1-41 $\dfrac{d^2x}{dt^2} + 5.831\dfrac{dx}{dt} + 6.25x = 0$

3. The auxiliary equations for these three cases are
   Case A   [_____]
   Case B   [_____]
   Case C   [_____]

- - - - - - - - - - - - - - - -

Case A $m^2 + 3m + 6.25 = 0$
Case B $m^2 + 5m + 6.25 = 0$
Case C $m^2 + 5.831m + 6.25 = 0$

   The roots of the auxiliary equation in each of these cases are
   Case A $m_1 = [$_____$]$; $m_2 = [$_____$]$
   Case B $m_1 = [$_____$]$; $m_2 = [$_____$]$
   Case C $m_1 = [$_____$]$; $m_2 = [$_____$]$

- - - - - - - - - - - - - - - -

Case A $m_1 = -1.5 + 2i$; $m_2 = -1.5 - 2i$
Case B $m_1 = -2.5$; $m_2 = -2.5$
Case C $m_1 = -1.4155$; $m_2 = -4.4155$

---

4. The solutions are (before the coefficients are determined):
   Case A $x = [$_____$]$
   Case B $x = [$_____$]$
   Case C $x = [$_____$]$

- - - - - - - - - - - - - - - -

Case A $x = e^{-1.5t}(k_1 \cos 2t + k_2 \sin 2t)$ (1-39a)
Case B $x = e^{-2.5t}(k_1 + k_2 t)$ (1-40a)
Case C $x = k_1 e^{-1.42t} + k_2 e^{-4.42t}$ (1-41a)

5. If we have the same initial conditions as in the preceding example,
i.e., $t = 0$; $x = x_1$; $dx/dt = 0$; the solutions are

Case A $x = e^{-1.5t}$ [_____] $\cos 2t$ + [_____] $\sin 2t$

— — — — — — — — — — — — — — — — — —

$k_1 = x_1$ $k_2 = 0$

_____

Case B $x = e^{-2.5t}$ [_____] + [_____] t

— — — — — — — — — — — — — — — — — —

$k_1 = x_1$ $k_2 = 2.5x_1$

_____

Case C $x = $ [_____] $e^{-1.42t}$ + [_____] $e^{-4.42t}$

— — — — — — — — — — — — — — — — — —

$k_1 = 1.473x_1$ $k_2 = -0.473x_1$

_____

6. In Case A, the motion is oscillatory, gradually dying out. In Cases
B and C, the motion of the mass is also oscillatory. [true, false]

— — — — — — — — — — — — — — — — — —

false

_____

In all three cases, the mass eventually settles (at $t \to \infty$), into the
equilibrium position of $x =$ [_____]

— — — — — — — — — — — — — — — — — —

zero

_____

In Case A, the mass initially passes through $x = 0$ and overshoots
this position, compressing the spring. [true, false]

— — — — — — — — — — — — — — — — — —

true

_____

It also does this in Case B. [true, false]

— — — — — — — — — — — — — — — — — —

false

_____

Reference Equations

1-39 $\dfrac{d^2x}{dt^2} + 3\dfrac{dx}{dt} + 6.25x = 0$

1-40 $\dfrac{d^2x}{dt^2} + 5\dfrac{dx}{dt} + 6.25x = 0$

1-41 $\dfrac{d^2x}{dt^2} + 5.831\dfrac{dx}{dt} + 6.25x = 0$

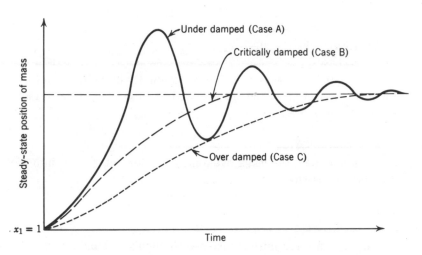

FIGURE 1-4   (The figures are drawn for $x_1 = 1$).

7. *Comments.* Case B, incidentally, is given a special name because of the repeated roots. If k_f had been smaller in magnitude than 5 newtons/meter/sec by even a minute amount, the solution would have been oscillatory. The friction force k_f (proportional to velocity) is termed a *damping force* and the critical value of k_f, which yields repeated roots, is termed the *critical damping case* [Ref. Eqs. (1-39), (1-40), (1-41)].

Case C is also given a name. In this case, the mass-spring system is said to be *"overdamped."* The overdamped system is slow acting compared to the *"under-damped"* system but is nonoscillatory. A graphical picture will bring out the nature of the three cases more clearly (see Fig. 1-4).

Evidently Case A would be termed [＿＿＿-damped].

- - - - - - - - - - - - - - - - - - -

underdamped

SECTION 4: A SIMPLE TRANSIENT ELECTRIC CIRCUIT PROBLEM

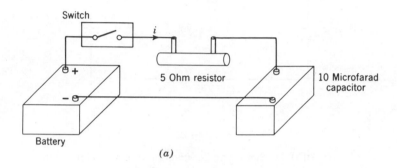

FIGURE 1-5a   Sketch of a circuit containing a battery,
a switch, resistor, and a capacitor.

1. Let us take up the problem of calculating the current i, in the circuit
of Fig. 1-5a. The problem of determining the equivalent circuit for
the actual circuit of Fig. 1-5a is not part of our problem here, since
this step requires the application of electric circuit theory. We will
take the equivalent circuit of Fig. 1-5b as given. A differential
equation model of this circuit is obtained by applying Kirchhoff's
Laws, specifically Kirchhoff's Voltage Law, since there is only one
current loop. The voltage drops in the direction of the current flow
are
Battery [_____ drop]

$-$ $-$ $-$ $-$ $-$ $-$ $-$ $-$ $-$ $-$ $-$ $-$ $-$ $-$ $-$ $-$ $-$ $-$

-6 volts

Resistor [____]

$-$ $-$ $-$ $-$ $-$ $-$ $-$ $-$ $-$ $-$ $-$ $-$ $-$ $-$ $-$ $-$ $-$ $-$

$5i$

(continued next page)

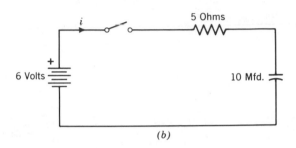

(b)

FIGURE 1-5b Schematic Diagram of an equivalent circuit
for the circuit of figure 1-5a.

Capacitor [_____]
- - - - - - - - - - - - - - - - - -

$\int_0^t \dfrac{i\ dt}{10 \times 10^{-6}} + (V_c)_{t=0}$

The resulting voltage equation is [_____]
- - - - - - - - - - - - - - - - - -

$-6 + 5i + \int_0^t 10^5 i\ dt + (V_c)_{t=0} = 0$ (1-42)

2. We now have Eq. (1-42), which does not look like a differential equation. Our first step in solution is to put it in differential equation
form. To do this we operate on this equation to remove the integral
sign. This is done by [_____
_____].
- - - - - - - - - - - - - - - - - -

differentiating all terms with respect to time, t.

This yields the differential equation [_____].
- - - - - - - - - - - - - - - - - -

$5\ di/dt + 1 \times 10^{+5}\ i = 0$ (1-43)

Reference Equations

1-42 $-6 + 5i + \int_0^t 10^5 \; i \; dt + (V_c)_{t=0} = 0$

1-43 $5 \; di/dt + 1 \times 10^{+5} \; i = 0$

3. Equation (1-43) is a homogeneous, ordinary, linear differential equation with constant coefficients. The auxiliary equation is
[_____]

- - - - - - - - - - - - - - - - - -

$5m + 1 \times 10^5 = 0$

and the root is [_____].

- - - - - - - - - - - - - - - - - -

$m = -2 \times 10^4$

The solution of Eq. (1-43), without evaluating the coefficient, is
$i = [$_____$]$.

- - - - - - - - - - - - - - - - - -

$i = ke^{-2 \times 10^4 t}$ $\hspace{3cm}$ (1-45)

4. We are now ready to use the initial condition to determine the coefficient k in Eq. (1-45). First we have to decide what is the initial condition? The initial condition we need is $t = 0$ [__ = ?].

- - - - - - - - - - - - - - - - - -

$i = ?$

To determine this initial condition, we look at Eq. (1-42). For $t = 0$, this equation becomes [_____].

- - - - - - - - - - - - - - - - - -

$-6 + 5(i)_{t=0} + (V_c)_{t=0} = 0$ $\hspace{2.5cm}$ (1-46)

We can solve Eq. (1-46) for $(i)_{t=0}$, if we know $(V_c)_{t=0}$, the initial voltage to which the capacitor is charged.

> **Reference Equations**
>
> **1-42** $-6 + 5i + \int_0^t 10^5\ i\ dt + (V_c)_{t=0} = 0$
>
> **1-45** $i = k\ e^{-2} \times 10^4 t$
>
> **1-46** $-6 + 5(i)_{t=0} + (V_c)_{t=0} = 0$

5. The initial voltage $(V_c)_{t=0}$ is not given in the problem statement.
We need to specify a value in order to complete the solution. Let
us assume that there is no initial voltage on the capacitor; this
would be the actual case if the capacitor has not been charged for
a long time and any charge present had "leaked" away, or if the
capacitor had been deliberately discharged, or had never been
charged. Then from Eq. (1-46) we find, at $t = 0$ [$i =$ _____].

‒ ‒ ‒ ‒ ‒ ‒ ‒ ‒ ‒ ‒ ‒ ‒ ‒ ‒ ‒ ‒ ‒

$i = 1.2$ amp   (1-47)

The corresponding value of the coefficient k is $k =$ [_____].

‒ ‒ ‒ ‒ ‒ ‒ ‒ ‒ ‒ ‒ ‒ ‒ ‒ ‒ ‒ ‒ ‒

$k = 1.2$

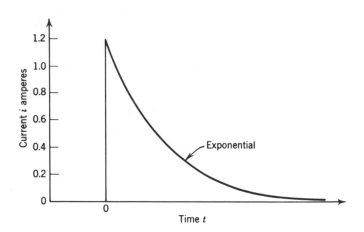

FIGURE 1-6 Sketch of the behavior of the current in the circuit of figure
1-5 in the case where there is no initial charge on the capacitor.

6. The complete solution of the problem is $i =$ [_____].

‒ ‒ ‒ ‒ ‒ ‒ ‒ ‒ ‒ ‒ ‒ ‒ ‒ ‒ ‒ ‒

$i = 1.2e^{-2} \times 10^4 t$

The form of this solution is shown in Fig. 1-6.

7. *Summary.* This closes our consideration, for the time being, of *homogeneous, linear differential equations with constant coefficients.* At this point, you should know how to:
   (1) Find the general solution of the homogeneous equation.
   (2) Given the initial conditions of the problem, find the values of the coefficients.

   As a scientist or engineer, of course, you will have to develop a capacity to study physical situations and determine the mathematical models, comprising both the differential equations *and* the initial conditions. These steps in modeling, however, are not matters of solving differential equations and will not be stressed further in this text.

PRACTICE PROBLEMS FOR CHAPTER I

1. The differential equation for a particular electric circuit involving a battery, two capacitors, and two resistors is

$$\frac{d^2i}{dt^2} + 175 \frac{di}{dt} + 2.5 \times 10^3 i = 0$$

with initial conditions of $i = 0.75$ amp and $di/dt = -625$ amp/sec at $t = 0$. Your solution for i is: $i = [\underline{\hspace{4cm}}]$.

- - - - - - - - - - - - - - - - -

$i = 3.52e^{-15.69t} + 4.27e^{-159.31t}$

2. The equation of motion of a particular rigid body in a system of bodies with connecting springs is

$$\frac{d^3x}{dt^3} + 60 \frac{d^2x}{dt^2} + 1100 \frac{dx}{dt} + 6000x = 0$$

with initial conditions at $t = 0$, $x = 0$, $dx/dt = 0$, $d^2x/dt^2 = 1000$ ft/sec^2. It is known that one term in the solution is $x = 5e^{-10t}$. Your solution for x is:

$x = 5e^{-10t} + [\underline{\hspace{5cm}}]$.

- - - - - - - - - - - - - - - - -

$x = 5e^{-10t} - 10e^{-20t} + 5e^{-30t}$

3. The equation for the current flowing into a 100 mfd. capacitor charged from a 50-volt source through a 100-ohm resistor has been found to be

$$100\,i + \int_0^t \frac{i\,dt}{1 \times 10^{-4}} = 50$$

In addition, it is known that at $t = 0$, $i = 0.5$ and $di/dt = 50$ amp/sec. Although the above is not quite an OLDECC, it can simply be converted into one and solved.
The solution is $i = [\underline{\hspace{3cm}}]$.

- - - - - - - - - - - - - - - - -

$i = 0.5e^{-100t}$

Any difficulties? Differentiate the original equation with respect to t to obtain the homogeneous OLDECC:

$$100\,\frac{di}{dt} + 10^4 i = 0$$

or

$$\frac{di}{dt} + 100\,i = 0$$

Also, note that although *two* initial conditions are given, only one is needed.

4. The differential equation for the displacement x of one mass in a
system of masses in a particular configuration as a function of time
is given by the equation

$$10\frac{d^3x}{dt} + 5000\frac{d^2x}{dt^2} + 1.1 \times 10^6 \frac{dx}{dt} + 1.5 \times 10^8 x = 0$$

The initial conditions are $t = 0$, $x = 0$, $dx/dt = 0$, $d^2x/dt^2 = 10^5$.
The solution is known to contain as one term:

$$x = 1.25e^{-100t} \sin 200t$$

Your complete solution is

x = [_____].

- - - - - - - - - - - - - - - - - - - -

$1.25e^{-300t} + e^{-100t} (1.25 \sin 200t - 1.25 \cos 200t)$

Any difficulty? It is clear from the one term in the solution that one
root of the auxiliary equation

$$10m^3 + 5000m^2 + 1.1 \times 10^6 m + 1.5 \times 10^8 = 0$$

or

$$m^3 + 500m^2 + 1.1 \times 10^5 m + 1.5 \times 10^7 = 0$$

is $m = -100 + j\,200$ ($j = \sqrt{-1}$). Hence one other root is $m = -100$
$- j200$, since complex roots must occur in conjugate pairs. The remaining root may be found by dividing the auxiliary equation by
$(m + 100 - j200)(m + 100 + j200) = m^2 + 200m + 50{,}000$, to obtain
the first order equation, $m + 300 = 0$.
Thus, the third root is $m = -300$.

5. The circuit equations for the currents i_1 and i_2 in two particular identically tuned coupled electric circuits of negligible losses, when one circuit is connected to a d-c source, are given by the equations

$$2\frac{di_1}{dt} + 300 \int_0^t i_1 dt + \frac{di_2}{dt} = 10$$

$$\frac{di_1}{dt} + 300 \int_0^t i_2 dt + 2\frac{di_2}{dt} = 0$$

The initial conditions are known to be at $t = 0$:

$$i_1 = 0 \qquad\qquad i_2 = 0$$

$$\frac{di_1}{dt} = \frac{20}{3} \qquad\qquad \frac{di_2}{dt} = \frac{10}{3}$$

$$\frac{d^2i_1}{dt^2} = 0 \qquad\qquad \frac{d^2i_2}{dt^2} = 0$$

$$\frac{d^3i_1}{dt^3} = -1000 \qquad \frac{d^3i_2}{dt^3} = +1000$$

Your solutions for the currents i_1 and i_2 as functions of time, are

$i_1 = [$————————————————————$]$.

$i_2 = [$————————————————————$]$.

- - - - - - - - - - - - - - - - - - -

$i_1 = 0.25 \sin 10t + 0.241 \sin 10\sqrt{3}\,t$

$i_2 = 1.0 \ \ \sin 10t - 0.385 \sin 10\sqrt{3}\,t$

Any difficulties? To solve, first differentiate both equations with respect to time to obtain

$$2\frac{d^2i_1}{dt^2} + 300i_1 + \frac{d^2i_2}{dt^2} = 0 \tag{1}$$

$$\frac{d^2i_1}{dt^2} + 300i_2 + 2\frac{d^2i_2}{dt^2} = 0 \tag{2}$$

The result is two simultaneous equations in i_1 and i_2 and their second derivatives. These equations must first be manipulated to obtain two equations, each of which is in *one* dependent variable only. This can be done, for example, by solving (2) for d^2i_1/dt^2:

$$\frac{d^2i_1}{dt^2} = 300i_2 - 2\frac{d^2i_2}{dt^2} \tag{3}$$

(continued next page)

and substituting this in (1) to obtain

$$- 600 i_2 - 4 \frac{d^2 i_2}{dt^2} + 300 i_1 + \frac{d^2 i_2}{dt^2} = 0$$

or

$$- 600 i_2 - 3 \frac{d^2 i_2}{dt^2} + 300 i_1 = 0 \qquad (4)$$

Equation (4) still has a term in i_1, which must be eliminated. This is done by differentiating (4) twice with respect to time.

$$- 600 \frac{d^2 i_2}{dt^2} - 3 \frac{d^4 i_2}{dt^4} + 300 \frac{d^2 i_1}{dt^2} = 0$$

and again substituting for $d^2 i_1 / dt^2$, using (3), to obtain

$$- 600 \frac{d^2 i_2}{dt^2} - 3 \frac{d^4 i_2}{dt^4} - 9 \times 10^4 i_2 - 600 \frac{d^2 i_2}{dt^2} = 0$$

Collecting terms, and dividing coefficients, the result is

$$\frac{d^4 i_2}{dt^4} + 400 \frac{d^2 i_2}{dt^2} + 30{,}000 i_2 = 0$$

The auxiliary equation is

$$m^4 + 400 m^2 + 30{,}000 = 0$$

This is the "biquadratic" case of the quartic equation. We first solve for m^2

$$m^2 = \frac{400 \pm \sqrt{16 \times 10^4 - 12 \times 10^4}}{2}$$

so that $m^2 = - 100$ or $- 300$.
Then, the four roots are

$$m = + \sqrt{-100}, \quad -\sqrt{-100}, \quad +\sqrt{-300}, \quad -\sqrt{-300}$$

or

$$m = + \sqrt{100}\, i, \quad -\sqrt{100}\, i, \quad +\sqrt{300}\, i, \quad -\sqrt{300}\, i.$$

The solution for i_2, then, with initial condition constants undetermined, is

$$i_2 = A \, \sin \sqrt{100}\, t + B \, \cos \sqrt{100}\, t + C \, \sin \sqrt{300}\, t + D \, \cos \sqrt{300}\, t$$

The remainder of the solution is straightforward.

Chapter II

SOLUTIONS OF INHOMOGENEOUS LINEAR *NTH* ORDER (OLDECC) ORDINARY DIFFERENTIAL EQUATIONS WITH CONSTANT COEFFICIENTS

Introduction

Up to this point, we have learned how to:

1. Find the general solution of the homogeneous OLDECC.
2. Given the initial conditions of the problem find the values of the coefficients (in the general solution) for the particular case under consideration.

We now start looking into the solutions of inhomogeneous equations.

Part A—General Form of Inhomogeneous Equations

The general form of equation to be considered is

$$\frac{ad^{m-1}y}{dt^{m-1}} + \frac{bd^{m-2}y}{dt^{m-2}} + \ldots + my = f(t)$$

In the preceding portions of this work, we have considered exactly the same form of equations except that we have limited our previous consideration to cases in which, in place of $f(t)$, the right side of the equation is [_____].

- - - - - - - - - - - - - - - - - -

zero

We shall start our consideration of inhomogeneous OLDECC with some simple examples.

Part B—Some Elementary Examples

1. Let us consider the equation

$$\frac{d^2y}{dt^2} + 6\frac{dy}{dt} + 5y = 10 \tag{2-1}$$

Is $y = 2$ a solution of this equation? [_____]

- - - - - - - - - - - - - - - - - - -

yes

_____

Is there any other *constant* value of y which is a solution? [_____]

- - - - - - - - - - - - - - - - - - -

no

_____

Do you think that $y = 2$ is the whole solution of Eq. (2-1)? [_____]

- - - - - - - - - - - - - - - - - - -

no or maybe!

_____

If there is any *more to be included* in the solution of Eq. (2-1), this "more" *which is to make the solution complete* must comprise an added function of time. [_____]

- - - - - - - - - - - - - - - - - -

yes

_____

EXPLANATION: Let us write the solution of Eq. (2-1) in the form $y = K + f(t)$, in which K is a constant. To substitute this expression for y in Eq. (2-1), we need

$$\frac{d^2}{dt^2}[K + f(t)] = 0 + \frac{d^2f(t)}{dt}$$

$$\frac{d^2}{dt}[K + f(t)] = 0 + \frac{d\ f(t)}{dt}$$

the result of substitution of $y = K + f(t)$ in Eq. (2-1), then, is

$$\frac{d^2\ f(t)}{dt} + 6\frac{d\ f(t)}{dt} + 5[(K + f(t)] = 10 \tag{2-2}$$

If we apply the restriction that $f(t)$ contains no constant terms it is clear that Eq. (2-1) actually yields two equations:

$$5K = 10, \quad or \quad K = 2$$

and

$$\frac{d^2\ f(t)}{dt^2} + 6\frac{d\ f(t)}{dt} + 5\ f(t) = 0 \tag{2-3}$$

The first of the above is the constant "solution" for y that we have already found. The second equation specifies what $f(t)$ must be.

We will now consider the solution for $f(t)$ in Eq. (2-3).

Reference Equations

2-1 $\dfrac{d^2y}{dt^2} + 6\dfrac{dy}{dt} + 5y = 10$

2-3 $\dfrac{d^2 f(t)}{dt^2} + 6\dfrac{d f(t)}{dt} + 5 f(t) = 0$

---

2. Eq. (2-3) is of a type with which we are already familiar; it is a
   [_____] OLDECC.

   _ _ _ _ _ _ _ _ _ _ _ _ _ _ _ _ _ _ _

   homogeneous

We know how to solve such an equation from our study of Chapter I.
The solution is $f(t) = k_1$ [_____] $+ k_2$ [_____].

_ _ _ _ _ _ _ _ _ _ _ _ _ _ _ _ _ _ _

$f(t) = k_1 e^{-t} + k_2 e^{-5t}$

---

3. Thus the complete solution of Eq. (2-1), without actually evaluating
   the constant coefficients (determination of which requires some initial conditions of the problem) is y = [__] + [_____] + [_____].

   _ _ _ _ _ _ _ _ _ _ _ _ _ _ _ _ _ _ _

   $y = 2 + k_1 e^{-t} + k_2 e^{-5t}$

---

4. Let us consider another example:

$$\frac{d^2y}{dt^2} + 10y = 2490 \sin 50t \qquad (2\text{-}4)$$

Consider as a possible solution to Eq. (2-4) the function

$$y = A \cos 50t \qquad (2\text{-}5)$$

Can any value of A be chosen so that Eq. (2-5) is a solution of Eq. (2-4)? [_____]

‒ ‒ ‒ ‒ ‒ ‒ ‒ ‒ ‒ ‒ ‒ ‒ ‒ ‒ ‒ ‒ ‒ ‒

no

As a result of going through the process of trying $y = A \cos 50t$ as a solution, can you suggest a trigonometric function different from Eq. (2-5) that would be a solution of Eq. (2-4)? If so your suggestion is $y = [$_____$]$.

‒ ‒ ‒ ‒ ‒ ‒ ‒ ‒ ‒ ‒ ‒ ‒ ‒ ‒ ‒ ‒ ‒ ‒

$A \sin 50t$

EXPLANATION: $y = A \sin 50t$ is a solution of Eq. (2-4), if $A = -1$, as you can readily verify by substitution in the equation.

Now that we have determined a function that is a solution of Eq. (2-4), on which we are working, it is appropriate to inquire whether there is anything that should be added to this particular solution to obtain a "complete" solution. If there is to be anything additional, it can be either a constant or another function of time, or both. We will now investigate these possibilities.

Reference Equations

2-4 $\dfrac{d^2y}{dt^2} + 10y = 2490 \sin 50t$

---

5. First, is there any *constant* value of y, i.e., $y = c$, which satisfies Eq. (2-4)? [yes, no]

— — — — — — — — — — — — — — — — —

no

Next, we will determine whether there is any function of time, $f(t)$, in addition to the $y = -\sin 50t$ already found, that can be included in the solution of Eq. (2-4). If there is such an additional function $f(t)$, it must be such that the whole solution for $y(t)$ is [_____].

— — — — — — — — — — — — — — — — —

$y = -\sin 50t + f(t)$

Further, we can see that this function $f(t)$ must be the solution of the homogeneous equation [_____].

— — — — — — — — — — — — — — — — —

$$\dfrac{d^2y}{dt^2} + 10y = 0 \qquad\qquad (2\text{-}6)$$

We already know from Chapter I that the solution of the homogeneous equation (2-6) is $y = $ [_____].

— — — — — — — — — — — — — — — — —

$y = k_1 \sin 10t + k_2 \cos 10t$

(without evaluation of the coefficients, because initial conditions are not given).

---

6. Thus the *full* general solution of the differential equation (2-4) is $y = $ [_____].

— — — — — — — — — — — — — — — — —

$y = -\sin 50t + k_1 \sin 10t + k_2 \cos 10t$

If the differential equation (2-4) had been the result of a solution of an actual problem, all that would remain for a complete solution is the evaluation of the constants k_1 and k_2 which appear in the solution of the [_____] equation.

— — — — — — — — — — — — — — — — —

homogeneous

---

7. Let us take stock of what we have learned from the foregoing two examples. In an nth-order OLDECC of the form:

$$a \frac{d^{m-1}y}{dx^{m-2}} + b \frac{d^{m-2}y}{dx^{m-2}} + \cdots + m y = g(x) \qquad (2\text{-}7)$$

it appears that the complete solution comprises two parts:

(1) a particular solution, $y = h(x)$, which satisfies *this* equation plus

(2) the complete solution of the *homogeneous* equation

$$a \frac{d^{m-1}y}{dx^{m-1}} + b \frac{d^{m-2}y}{dx^{m-2}} + \cdots + m y = 0$$

obtained by replacing the function on the right-hand side of the equation by zero.

We have not proved in a general way that such a solution is proper, but this solution certainly emerged as appropriate in the two cases we considered. Of course, it must have seemed to you that we simply invented rather than logically derived the two *particular* functions that comprised the particular solution $y = h(x)$ in these cases. Before looking at the possibilities of logical derivation of such solutions we will discuss the general procedure for solution of the inhomogeneous OLDECC and give some definitions.

---

8. The function $g(x)$ in Eq. (2-7) is called the *driving function, forcing function,* or *driving force*.

The solution designated $y = h(x)$, above, is called the *particular solution,* or sometimes the *particular integral*. The solution of the *homogeneous* equation, obtained by replacing the driving function by zero, is called the *complementary solution* or *complementary function*.

---

9. Thus the whole solution of an nth-order linear differential equation with constant coefficients of the form:

$$a\frac{d^m y}{dt^m} + b\frac{d^{m-1}y}{dt^{m-1}} + \cdots + n\,y = g(t) \qquad (2\text{-}8)$$

would appear to comprise the sum of two terms: [_____] solution + [_____] solution.

----

complementary   particular

The *particular solution* of (2-8) is any function $y = h(t)$ that we can find that satisfies Eq. (2-8). The *complementary solution* of (2-8) is the complete solution of the *homogeneous equation* that we obtain by replacing the right-hand side of (2-8) by [_____].

----

zero

The complementary solution of Eq. (2-8) will contain [$m + 1$, m, $m - 1$] undetermined coefficients.

----

m

These coefficients must be evaluated by reference to the particular conditions of the problem situation (for which the solution of the differential equation is supposed to be a model).

The foregoing procedure is, in fact, the general method of determining solutions of the class of differential equations treated in this work.

Incidentally, if we have found a function is a particular solution, we can say that it is *the* particular solution. There is no need to seek another, there is only *one* particular solution possible in each case.

We need, however, some more specific directions for finding the particular solution portions of the whole solution. These methods are discussed next.

Part C—Finding the Particular Solution of
Inhomogeneous OLDECC

1. We have some discouraging news for you. There is no *easy* general direct method for finding the particular solutions of the inhomogeneous, linear, ordinary differential equations with constant coefficients. However, some experience will show the student that if he is resourceful, he can readily invent or otherwise devise the particular solutions required in any case that he is likely to encounter in dealing with physical situations.

The driving function is actually the physical force or agent that acts upon the physical configuration from which the OLDECC model arises. In the problems discussed in Chapter I, the configuration had no external force acting upon them. They were merely disturbed, as by pulling a mass from its rest position so as to stretch a spring and then releasing it. If the mass-spring configuration of Chapter I had been subjected to a steady disturbing force, or a pulsating force, we would have had an inhomogeneous rather than homogeneous OLDECC. This short explanation indicates how driving functions arise and why the kinds of driving function we can expect to arise in real problems are not very great in number.

Can you think of some forms of driving function you would expect? The simplest is one we have already considered, a driving function which is a [_____].

- -

constant

Some other possible driving functions are sinusoidal functions of time, or combinations of such functions, exponential functions or combinations of exponentials, and polynomial algebraic expressions:

Trigonometric Driving Function $f(t) = A_1 \sin k_1 t + B_1 \cos k_1 t \ldots$
Exponential Driving Function $f(t) = A_1 e^{k_1 t} + A_2 e^{k_2 t} + A_3 e^{k_3 t} + \ldots$
Polynomial Driving Function $f(t) = A + Bt + Ct^2 + Dt^3 + \ldots$

These examples do not seem to comprise a wide variety of possibilities; however, such functions as $\log k_1 t$, $\sec k_1 t$, etc., are probably excluded from arising in most practical problems because they are not well defined for all finite values of t in the range from zero upward and we expect, naturally, that any equation that describes a physical situation will yield physically meaningful results, an expectation that excludes undefined or infinite values of the dependent variable, or the physical phenomenon it represents.

Section 1 CONSTANT DRIVING FUNCTION

1. Let us consider the problem of solving the inhomogeneous OLDECC:

$$5\frac{d^2y}{dx^2} + 6\frac{dy}{dx} + 15y = 90 \qquad (2\text{-}9)$$

To solve this we need to find (*a*) the [_____] solution

— — — — — — — — — — — — — — — — —

particular

---

and (*b*) the [_____] solution.

— — — — — — — — — — — — — — — — —

complementary

From our previous experience, we expect that the particular solution will be of the form $y = $ [_____].

— — — — — — — — — — — — — — — — —

$y = $ constant (call it *A*)

If $y = A$, the derivative $dy/dx = $ [_____] and the next derivative $d^2y/dx^2 = $ [_____]

— — — — — — — — — — — — — — — —

zero   zero

---

so that, if we substitute $y = A$ in Eq. (2-9), we obtain $A = $ [____].

— — — — — — — — — — — — — — — — —

$15A = 90$ or $A = 6$

This is the particular solution for Eq. (2-9).

2. Generalizing from this experience, we can see that in the nth-order inhomogeneous OLDECC, of the form:

$$a_0 \frac{d^n y}{dt^n} + a_1 \frac{d^{n-1}y}{dt^{n-1}} + \cdots + a_n y = C \qquad (2\text{-}10)$$

(C, a_i's are constants)

the particular solution will be $y = [\underline{\hspace{2cm}}]$.

_ _ _ _ _ _ _ _ _ _ _ _ _ _ _ _ _ _

$y = C/a_n$

In this case, we obviously have a ready and simple method for finding the particular integral (solution). As far as the whole solution is concerned, we already know how to find the complementary solution by the methods discussed in Chapter I.

Without stopping to formally complete the solution of Eqs. (2-9) and (2-10), we will proceed with a more complicated driving function.

**Section 2   SINUSOIDAL DRIVING FUNCTIONS:
A SIMPLE EXAMPLE**

1. As an example of a differential equation that might describe a linear physical situation in which there is a sinusoidal driving function, consider the equation

$$0.25 \frac{d^2 y}{dt^2} + 2 \frac{dy}{dt} + 10y = 50 \cos 4t \qquad (2\text{-}11)$$

Let us inquire into the *particular solution* of this equation. We will proceed by "guessing" at a solution for y and substituting it into Eq. (2-11). If the guess is correct, it will make the left-hand side of the equation equal to the right-hand side *for all values of the independent variable, t.*

Reference Equation

2-11 $0.25 \dfrac{d^2y}{dt^2} + 2 \dfrac{dy}{dt} + 10y = 50 \cos 4t$

2. First, is there any value of a constant, C_1, such that $y = C_1$ is the particular solution of Eq. (2-11)? [yes, no].
If so, the value of C_1 is [_____].

- - - - - - - - - - - - - - - - - -

no, (there isn't any possible C)

EXPLANATION: This is the case because when we substitute
$y = C_1$ *in Eq. (2-11) we obtain* [_____].

- - - - - - - - - - - - - - - - - -

$10C_1 = 50 \cos 4t$ *(2-12)*

There is, therefore, no constant value of C_1 in (2-12) which will satisfy (2-11) for all values of t.

3. As a matter of fact, the only functions which can be substituted for y that are likely to produce functions which can be equal to the sinusoidal function on the right-hand side of Eq. (2-11) are themselves [_____] functions.

- - - - - - - - - - - - - - - - - -

sinusoidal

Moreover, these sinusoidal functions should have the argument [__].

- - - - - - - - - - - - - - - - - -

$4t$

As a possible solution, consider
$$y = C_1 \cos 4t \qquad (2\text{-}13)$$
If a suitable value of the constant C_1 is chosen, is Eq. (2-13) the particular solution of (2-11)? [_____]

- - - - - - - - - - - - - - - - - -

no

EXPLANATION: $y = C_1 \cos 4t$ is not the particular solution of Eq. (2-11), because the substitution of $y = C_1 \cos 4t$, $dy/dt = -4C_1 \sin 4t$, $d^2y/dt^2 = -16C_1 \cos 4t$ in Eq. (2-11) yields the equation,
$$-4C_1 \cos 4t - 8C_1 \sin 4t + 10C_1 \cos 4t = 50 \cos 4t \quad \text{(2-14)}$$
and there is no single value of C_1 which will make the left-hand side of (2-14) equal to the right-hand side for all values of t.

4. If a suitable value of C_2 is chosen, is $y = C_2 \sin 4t$ the particular solution of (2-11)? [_____]

- - - - - - - - - - - - - - - - - - - -

no

___

Quite clearly, similar difficulties arise in using either $y = C_1 \cos 4t$ or $y_2 = C_2 \sin 4t$ as a particular solution of Eq. (2-11).

However, our results suggest that we should try as the particular solution of Eq. (2-11) $y = [$_____$] + [$_____$]$.

- - - - - - - - - - - - - - - - - - - -

$y = C_1 \cos 4t + C_2 \sin 4t$ (2-15)

___

In fact, with suitable values of the constants C_1 and C_2 is the function, Eq. (2-15), the particular solution of Eq. (2-11)? [_____]

- - - - - - - - - - - - - - - - - - - -

yes

___

EXPLANATION: This is the case because the substitution of
$y = C_1 \cos 4t + C_2 \sin 4t; \; dy/dt = -4C_1 \sin 4t + 4C_2 \cos 4t;$
$d^2y/dt^2 = -16C_1 \cos 4t - 16C_2 \sin 4t$ *in (2-11) yields*
$[$_____

_____$] = 50 \cos 4t.$

- - - - - - - - - - - - - - - - - - - -

$-4C_1 \cos 4t - 4C_2 \sin 4t - 8C_1 \sin 4t + 8C_2 \cos 4t + 10C_1 \cos 4t + 10C_2 \sin 4t = 50 \cos 4t$ (2-16)

___

In order to satisfy Eq. (2-16) for all values of t the coefficients of like terms on the left-hand side of Eq. (2-16) should be equal to the coefficients of like terms on the right-hand side.
Thus we find $-4C_1 + 8C_2 + 10C_1 = [$___$]$

- - - - - - - - - - - - - - - - - - - -

50

___

and $-4C_2 - 8C_1 + 10C_2 = [$___$]$

- - - - - - - - - - - - - - - - - - - -

zero

___

or $6C_1 + 8C_2 = 50; \; 6C_2 - 8C_1 = 0$
of which, the solutions are $C_1 = [$___$], C_2 = [$___$]$

- - - - - - - - - - - - - - - - - - - -

3   4

___

so that the particular solution of Eq. (2-11) is the function
$y = [$_____$].$

- - - - - - - - - - - - - - - - - - - -

$y = 3 \cos 4t + 4 \sin 4t$

5. Hence the complete solution of Eq. (2-11), which includes, in addition to the particular solution, the complementary function, is

$$y = [\underline{\hspace{6cm}}].$$

- - - - - - - - - - - - - - - - - -

$$y = 3 \cos 4t + 4 \sin 4t + e^{-4t} (k_1 \cos \sqrt{24}\, t + k_2 \sin \sqrt{24}\, t)$$

with constants k_1 and k_2 still to be determined from the initial conditions of the physical situation from which the original differential equation arises.

Section 3   SINUSOIDAL DRIVING FUNCTIONS:
A FIRST-TRIAL GENERALIZATION

1. As a result of our experience with our simple example, let us suggest the following generalizations:

In an inhomogeneous OLDECC of the form

$$a_0 \frac{d^n y}{dt^n} + a_1 \frac{d^{n-1}y}{dt^{n-1}} + \cdots + a_n y = f(t) \tag{2-17}$$

in which the driving function $f(t)$ comprises the sum of one or more sinusoidal terms such as $A_1 \cos w_1 t + B_1 \sin w_1 t + A_2 \cos w_2 t + B_2 \sin w_2 t + \ldots$ (in which A_1, B_1, A_2, B_2, etc., are constants), the particular solution for y will comprise terms of the form:
   $y = C_1 \cos w_1 t + C_2 \sin w_1 t + C_3 \cos w_2 t + C_4 \sin w_2 t + \ldots$ *in which the values C_1, C_2, C_3, C_4, ... etc., will be determined by* (select the correct one or more of the following):

2. (a) Substitution of the particular solution into the whole differential equation (2-17) and choosing values of the constants C_1, C_2, C_3, C_4, etc., which will satisfy the whole equation for *all* values of the independent variable.

   (b) Substitution of the particular solution into the *homogeneous* equation (i.e., equation obtained when $f(t)$ in (2-17) is replaced by zero) and choosing values of the constants C_1, C_2, C_3, C_4, etc., which will satisfy the homogeneous equation for all values of the independent variable.

   (c) Substitution of the initial conditions of the problem for which the differential equation is derived.

- - - - - - - - - - - - - - - - - -

(a) above is correct; (b) and (c) are not correct

This generalization certainly correctly describes the procedure that worked in the above cases. Before accepting this generalization as valid, however, we should test it on a few more specific examples.

Section 4 SINUSOIDAL DRIVING FUNCTIONS:
ANOTHER (AND IMPORTANTLY DIFFERENT) EXAMPLE

1. Let us test the foregoing first generalization (as to the particular
solutions in the case of sinusoidal driving functions) upon the following example:

$$\frac{d^2y}{dt^2} + 9y = 101.4 \sin 3t \qquad (2\text{-}18)$$

According to our preceding generalization, we would expect the
particular solution to comprise:

$y = [\underline{\hspace{2cm}}] + [\underline{\hspace{1.5cm}}]$

- - - - - - - - - - - - - - - -

$$y = A \sin 3t + B \cos 3t \qquad (2\text{-}19)$$

___

We can find A and B by determining $dy/dt = [\underline{\hspace{1.5cm}}] +$
$[\underline{\hspace{1.5cm}}]$.

- - - - - - - - - - - - - - - -

$dy/dt = 3 \cos 3t - 3B \sin 3t$

___

and $d^2y/dt^2 = [\underline{\hspace{1.5cm}}] + [\underline{\hspace{1.5cm}}]$.

- - - - - - - - - - - - - - - -

$d^2y/dt^2 = -9A \sin 3t - 9B \cos 3t$

___

When these values are substituted in Eq. (2-18), the left-hand
side of Eq. (2-18) becomes

$[\underline{\hspace{5cm}}] = [\underline{\hspace{1.5cm}}]$

- - - - - - - - - - - - - - - -

$$\sin 3t\,(9A - 9A) + \cos 3t\,(9B - 9B) = 0 \qquad (2\text{-}20)$$

Reference Equations

2-18 $\dfrac{d^2y}{dt^2} + 9y = 101.4 \sin 3t$

2-19 $y = A \sin 3t + B \cos 3t$

2-20 $\sin 3t (9A - 9A) + \cos 3t (9B - 9B) = 0$

2. Clearly, there are no values of A and B that will make Eq. (2-20) equal to 101.4 sin 3t. Therefore our generalization of the preceding section does not work for finding the particular solution of Eq. (2-18). Before throwing away this generalization, however, we should at least seek the reason for its failure.

The solution (2-19) fails for Eq. (2-18) because the roots of the auxiliary equation are $m =$ [_____], [_____].

- - - - - - - - - - - - - - - - - -

$m = + 3i, \quad m = - 3i$

_____

These correspond to a complementary solution of the form:
$y =$ [_____] + [_____]

- - - - - - - - - - - - - - - - - -

$y = k_1 e^{3it} + k_2 e^{-3it}$

_____

or, using trigonometric identities, a complementary solution of the form: $y =$ [_____] + [_____].

- - - - - - - - - - - - - - - - - -

$y = k_1 \sin 3t + k_2 \cos 3t$ (2-21)

_____

The arguments, 3t, of the trigonometric functions in the *comple-mentary* solution (2-21) are the same as the argument of the *driving function* in Eq. (2-18), 3t. This is the source of the trouble.

3. We know that the substitution in the homogeneous equation of any function which is a root of the auxiliary equation will make the homogeneous equation equal to zero. If it is zero it cannot be equal to a driving function $f(t)$. Hence whenever the driving function suggests a particular solution which is a root of the auxiliary equation, this particular solution is invalid and another must be sought. Thus we have one exception to the proposed particular solution for the inhomogeneous OLDECC in which the driving function is periodic. This occurs when the argument ω_i of any term in the driving function (which may comprise the sum of a number of different periodic functions) corresponds to a root of the auxiliary equation,

$m = [_____], [_____]$.

_ _ _ _ _ _ _ _ _ _ _ _ _ _ _ _ _ _ _

$m = i\omega_1, \ -i\omega_1$

In this case, what do we do about the particular solution? In the case in which some one term in the driving function has an argument $\omega_1 t$ identical with the root of the auxiliary equation, $m = \pm i\omega_1$, the particular solution then contains terms of the form:

$$y = At \sin \omega_1 t + Bt \cos \omega_1 t \qquad (2\text{-}22)$$

The validity of this solution (2-22) for the differential equation (2-18) may be demonstrated as follows:

> **Reference Equation**
>
> **2-18** $\dfrac{d^2y}{dt^2} + 9y = 101.4 \sin 3t$

4. Our solution for Eq. (2-18) is $y = [A(\underline{\hspace{2cm}}) + B(\underline{\hspace{2cm}})]$.

- - - - - - - - - - - - - - - - - -

$y = At \cos 3t + Bt \sin 3t$ (2-23)

_____

In order to check this solution, we need to find dy/dt and d^2y/dt^2. These follow: $dy/dt = [\underline{\hspace{4cm}}$
$\underline{\hspace{3cm}}]$.

- - - - - - - - - - - - - - - - - -

$dy/dt = A \cos 3t - 3At \sin 3t + B \sin 3t + 3Bt \cos 3t$

_____

and $d^2y/dt^2 = \cos 3t \, [\underline{\hspace{2cm}}] - \sin 3t \, [\underline{\hspace{2cm}}]$.

- - - - - - - - - - - - - - - - - -

$d^2y/dt^2 = \cos 3t \, (6B - 9At) - \sin 3t \, (6A + 9Bt)$

_____

When these values are substituted in Eq. (2-18), we obtain
$[\underline{\hspace{5cm}}] = 101.4 \sin 3t$

- - - - - - - - - - - - - - - - - -

$6B \cos 3t - 6A \sin 3t = 101.4 \sin 3t$ (2-24)

_____

In order for Eq. (2-24) to be correct, the coefficients A and B should be $A = [\underline{\hspace{2cm}}]$, $B = [\underline{\hspace{2cm}}]$.

- - - - - - - - - - - - - - - - - -

$A = -101.4/6, \quad B = 0$

_____

Hence, the function y given by Eq. (2-23) is a correct general form of solution for Eq. (2-18). In this particular case, of course, the constant B is zero.

We can now improve upon the generalization of Section 3 for the solution of the inhomogeneous OLDECC in which the driving function comprises one or more trigonometric functions.

Section 5   SINUSOIDAL DRIVING FUNCTIONS:
A SECOND (AND IMPROVED) GENERALIZATION

1. As a result of our experience in Sections 2, 3, and 4, we can now make an improved generalization for the solutions of the inhomogeneous OLDECC in which the driving function comprises one or more periodic terms.

In an inhomogeneous linear, ordinary differential equation with constant coefficients of the form:

$$a_0 \frac{d^n y}{dt^n} + a_1 \frac{d^{n-1} y}{dt^{n-1}} + \ldots + a_n y = f(t)$$

in which the driving function $f(t)$ comprises the sum of one or more periodic terms such as $A_1 \cos \omega_1 t + B_1 \sin \omega_1 t + A_2 \cos \omega_2 t + B_2 \sin \omega_2 t + \ldots$ *(in which, A_1, B_1, A_2, B_2, etc., are constants) the particular solution for y will comprise terms of the form:*

$y = C_1 \cos \omega_1 t + C_2 \sin \omega_1 t + C_3 \cos \omega_2 t + C_4 \sin \omega_2 t + \ldots$
unless one or more of the arguments $(\omega_1, \omega_2 \ldots)$ of the periodic terms correspond to roots $m = [\rule{1cm}{0.4pt}]$, $[\rule{1cm}{0.4pt}]$.

- - - - - - - - - - - - - - - - - -

$m = \pm i\omega_1, \ \pm i\omega_2$

---

of the auxiliary equation, $a_0 m^n + a_1 m^{n-1} + \ldots + a_n = 0$.

For each term in the driving function with an argument corresponding to one of these roots, ω_1, ω_2 the particular solution will contain terms of the form
$y = [\rule{1cm}{0.4pt}] \cos \omega_1 t + [\rule{1cm}{0.4pt}] \sin \omega_1 t + C_3 t \cos \omega_2 t + C_4 t \sin \omega_2 t$

- - - - - - - - - - - - - - - - - -

$C_1 t \cos \omega_1 t + C_2 t \sin \omega_1 t$

---

2. Incidentally, if we should happen to include *unnecessary* terms in a proposed particular solution by mistake, this will increase the algebraic work in substituting the solution in the equation but will introduce no errors. This is because the coefficients of any unnecessary terms will turn out to be equal to $[\rule{1cm}{0.4pt}]$ when the solution is substituted in the differential equation in order to evaluate the coefficients.

- - - - - - - - - - - - - - - - - -

zero

---

3. On the other hand, if we do not include all the necessary terms in a proposed (trial) particular solution, we will find it impossible to set the values of the coefficients so that the proposed particular solution satisfies the equation. [true, false]

----

true

---

4. We remind you that a correct *particular solution* of a linear differential equation with constant coefficients is unique, i.e., that there is only one such. Both the functions in the particular solution and the coefficients of these functions are uniquely determined by the differential equation alone and do not require additional physical information concerning the problem which is described by the differential equation.

Is this also true of

(a) the functions in the complementary solution: [yes, no]

----

yes

(b) the coefficients of these functions in the complementary solution: [yes, no]

----

no

EXPLANATION: You will remember, from Chapter I, that the functions in the complementary solution are uniquely determined by the differential equation, but their coefficients are determined by additional initial condition information not in the differential equation.

5. Before leaving this trigonometric form of driving functions, consider
the following examples and give suitable solutions. (The homogeneous part of the OLDECC will not be specified so that you cannot
evaluate the coefficients or eliminate unnecessary terms. The roots
of the auxiliary equation are given, however.)
Driving function (D.F.): $20 \cos \omega t$ $(m \neq i\omega)$; particular solution form:
[_____].

– – – – – – – – – – – – – – – – – –

$A \sin \omega t + B \cos \omega t$

D.F.: $20 \sin 5t$ $(m = \pm i5)$; solution [_____].

– – – – – – – – – – – – – – – – – –

$At \sin 5t + Bt \cos 5t$

D.F.: $5 \sin 3t + 10 \cos 6t$ $(m \neq \pm 3i, \pm 6i)$; solution
[_____].

– – – – – – – – – – – – – – – – – –

$A_1 \sin 3t + B_1 \cos 3t + A_2 \sin 6t + B_2 \cos 6t$

D.F.: $3 + 15 \sin 0.5t$ $(m = \pm 3i)$; solution [_____].

– – – – – – – – – – – – – – – – – –

$A + A_1 \sin 0.5t + B_1 \cos 0.5t$

---

6. The actual values of the various constants A_1, B_1, etc., in the foregoing examples cannot be determined because the *differential equations* for these examples have not been given. [true, false]

– – – – – – – – – – – – – – – – –

true

It is apparent that when the driving function is the sum of two or
more different types of driving functions, the particular solutions
comprise the sums of the particular solutions which would be obtained for each of the terms of the driving function separately.
[true, false]

– – – – – – – – – – – – – – – – –

true

This is the case because we are dealing with [*linear equations,
differential equations, first-order equations*].

– – – – – – – – – – – – – – – – –

linear equations

The *principle of super position* applies to such linear systems.

Section 6   SUMMARY OF WORK TO THIS POINT

We are now ready to consider some additional types of driving functions (also known as driving force or forcing function); solutions for exponential and polynomial driving functions will be considered in Chapter III.

At this point, you should know how to:

(1) Find the general solution of a given homogeneous OLDECC.

(2) Given the initial conditions of the problem, find the values of the coefficients in the homogeneous OLDECC.

from Chapter I

(3) Given an inhomogeneous OLDECC, find the complete solution comprising the complementary solution and the particular solution for the cases of a steady (constant) driving function and a sinusoidal driving function.

(4) Given the initial conditions of the problem find the value of the coefficients in the complementary solution.

from Chapter II

(5) Relate the complementary and particular solutions to transient and steady-state behavior of the problem configuration.

PRACTICE PROBLEMS FOR CHAPTER II

1. The differential equation for the current i_3 in a particular circuit consisting of one (lossy) inductor in series with another, which is also shunted with a resistor, and a 20-volt source is

$$\frac{d^2 i_3}{dt^2} + 280 \frac{di_3}{dt} + 12{,}800 = 6400$$

The initial conditions in this circuit are

$$t = 0 \quad i_3 = 0 \quad \frac{di_3}{dt} = 40$$

Your solution for i_3 is [_____
_____].

- - - - - - - - - - - - - - - - - -

$i_3 = 0.5 - 0.068e^{-222.5t} - 0.432e^{-57.45t}$

2. In the circuit of problem 1, the energy source is changed to a sinusoidal source of voltage, with this voltage applied to the circuit at $t = 0$. The differential equation for the current i_3 then becomes

$$\frac{d^2 i_3}{dt^2} + 280 \frac{di_3}{dt} + 12,800 = -2720 \sin 200t - 560 \cos 200t$$

The initial conditions are

$$t = 0 \quad i_3 = 0 \quad \frac{di_3}{dt} = 0$$

Your solution in this case is

$$i_3 = [\underline{\hspace{9cm}}].$$

$$i_3 = 0.0318e^{-22.5t} - 0.1318e^{-54.45t} + 0.1 \cos 200t$$

---

3. In the physical configuration described by the differential equation

$$1 \times 10^{-4} \frac{d^2 y}{dx^2} + 3 \times 10^{-2} \frac{dy}{dx} + 3.25y = 9 \cos 300t + 23.5 \sin 300t + 6.5$$

the initial conditions are

$$t = 0 \quad y = 2, \quad \frac{dy}{dt} = 3102$$

Your solution is

$$y = [\underline{\hspace{9cm}}].$$

$$y = e^{-150t} (\sin 100t) + 10 \sin 300t + 2$$

4. The equation for the displacement x of one of two masses in a particular coupled system of two masses and two linear springs, in a case in which the damping is negligibly small, and a sinusoidal time-varying force is applied, is given by

$$\frac{d^4x}{dt^4} + 0.29\frac{d^2x}{dt^2} + 0.01x = -0.21 \cos 0.5t$$

with initial conditions

$$t = 0 \quad x = 3, \quad \frac{dx}{dt} = -0.8, \quad \frac{d^2x}{dt^2} = 0.67, \quad \frac{d^3x}{dt^3} = 0.242$$

Your solution for x is

x = [_____].

- -

x = sin 0.2t + 2 cos 0.2t − 2 sin 0.5t + cos 0.5t + t sin 0.5t

---

EXPLANATION, if needed: This is a biquadratic case of a quartic equation. The auxiliary equation is

$$m^4 + 0.29m^2 + 0.01 = 0$$

with roots $m^2 = -0.04$ and $m^2 = -0.25$ or $m = +i\,0.2, -i\,0.2, +i\,0.5,$ and $-i\,0.5$. Since the argument of the driving function is also 0.5t, the particular integral will contain the form t sin 0.5t or t cos 0.5t or both.

Chapter III

PARTICULAR SOLUTION OF THE INHOMOGENEOUS OLDECC: SOME ADDITIONAL TYPES OF DRIVING FUNCTIONS AND SOME MORE GENERALIZATIONS

Section 1 SOLUTIONS FOR OLDECC WITH POLYNOMIAL DRIVING FUNCTIONS

1. Let us consider the following differential equation:

$$\frac{d^2x}{dt^2} + 4\frac{dx}{dt} + 3x = 2t - 3t^2 + 3t^3 \tag{3-1}$$

(Such a driving function as that in Eq. (3-1) might arise, for instance, as a result of a curve-fitting procedure in which a graphically described driving function is to be represented within a particular time interval by an approximation. This driving function, of course, becomes meaningless as $t \to \infty$, so that we would expect the physical situation represented to be one in which we were interested only in the behavior of x for values of t in a particular range of values, probably starting at $t = 0$.)

Reference Equation

3-1 $\dfrac{d^2x}{dt^2} + 4\dfrac{dx}{dt} + 3x = 2t - 3t^2 + 3t^3$

---

2. The particular solution must comprise some terms which, when substituted on the left-hand side of (3-1), will result in terms such as
$2t$, $3t^2$, and $3t^3$. Which of the following appear to qualify as at least
part of the particular solution? Please answer yes or no to each one.
(In the terms suggested, *A, B, C, D, E,* etc., are constants.)
A [_____]

- - - - - - - - - - - - - - - - - - - -

yes

Bt [_____]

- - - - - - - - - - - - - - - - - - - -

yes

Ct^2 [_____]

- - - - - - - - - - - - - - - - - - - -

yes

Dt^3 [_____]

- - - - - - - - - - - - - - - - - - - -

yes

Et^4 [_____]

- - - - - - - - - - - - - - - - - - - -

no

[_____] (anything you want to add)

- - - - - - - - - - - - - - - - - - - -

We can, of course, try a particular solution comprising all of the
above; if any term is unnecessary, its coefficients will turn out to
be [_____].

- - - - - - - - - - - - - - - - - - -

zero

If we have omitted a necessary term, it will be impossible to find
values of the coefficients (*A, B,* etc.) which will satisfy the equation. [correct, incorrect].

- - - - - - - - - - - - - - - - - - -

correct

---

3. For instance, let us test as a solution of (3-1) the function:

$$x = Bt + Ct^2 + Dt^3 \tag{3-2}$$

We first determine $\dfrac{dx}{dt} = [\underline{\hspace{3cm}}]$.

- - - - - - - - - - - - - - - - - -

$B + 2Ct + 3Dt^2$

---

and $\dfrac{d^2x}{dt^2} = [\underline{\hspace{3cm}}]$.

- - - - - - - - - - - - - - - - - -

$2C + 6\,Dt$

Substitution of the solution (3-2) in (3-1) yields $[\underline{\hspace{3cm}}$
$\underline{\hspace{6cm}}]$.

- - - - - - - - - - - - - - - - - -

$2C + 6Dt + 4B + 8Ct + 12Dt^2 + 3Bt + 3Ct^2 + 3Dt^3 = 2t - 3t^2 + 3t^3$

By equating the coefficients of like terms in t, we obtain

$2C + 4B = [\underline{\hspace{1.5cm}}]$ \hfill (3-3a)

- - - - - - - - - - - - - - - - - -

zero

$6D + 8C + 3B = [\underline{\hspace{1.5cm}}]$ \hfill (3-3b)

- - - - - - - - - - - - - - - - - -

2

$12D + 3C = [\underline{\hspace{1.5cm}}]$ \hfill (3-3c)

- - - - - - - - - - - - - - - - - -

-3

$[\underline{\hspace{3cm}}] = 3$ \hfill (3-3d)

- - - - - - - - - - - - - - - - - -

$3D = 3$

Reference Equations

3-1 $\dfrac{d^2x}{dt^2} + 4\dfrac{dx}{dt} + 3x = 2t - 3t^2 + 3t^3$

3-2 $x = Bt + Ct^2 + Dt^3$

---

4. The procedure yielded four *independent* equations (3-3*a*), (3-3*b*), (3-3*c*), and (3-3*d*) in three unknowns, *B*, *C*, and *D*. We cannot find any constants *B*, *C*, and *D* which will satisfy all four equations and, therefore, make (3-2) a particular solution of (3-1). This is because

(a) There were more terms than necessary in the proposed particular solution (3-2) [yes, no].

‒ ‒ ‒ ‒ ‒ ‒ ‒ ‒ ‒ ‒ ‒ ‒ ‒ ‒ ‒ ‒ ‒

no

(b) Some necessary term or terms in the particular solution were omitted. [yes, no]

‒ ‒ ‒ ‒ ‒ ‒ ‒ ‒ ‒ ‒ ‒ ‒ ‒ ‒ ‒ ‒ ‒

yes

Had there been simply too many terms in the proposed particular solution, the coefficient(s) of any unnecessary term(s) would have been [_____].

‒ ‒ ‒ ‒ ‒ ‒ ‒ ‒ ‒ ‒ ‒ ‒ ‒ ‒ ‒ ‒ ‒

zero

In this case, it would have been possible to find values of the remaining coefficients which would cause the proposed solution to satisfy the equation.

---

5. Let us try to decide what term or terms is missing. The missing term is probably [_____].

‒ ‒ ‒ ‒ ‒ ‒ ‒ ‒ ‒ ‒ ‒ ‒ ‒ ‒ ‒ ‒ ‒

a constant

---

6. To show that a constant, *A*, is all that is needed, we will try as a particular solution of Eq. (3-1) the function:

$$x = A + Bt + Ct^2 + Dt^3 \tag{3-3}$$

Reference Equation

3-3 $x = A + Bt + Ct^2 + Dt^3$

---

7. For this function, (3-3), we have $dx/dt = [\underline{\hspace{4cm}}]$.

- - - - - - - - - - - - - - - - - -

$B + 2Ct + 3Dt^2$

$d^2x/dt^2 = [\underline{\hspace{2cm}}]$.

- - - - - - - - - - - - - - - - - -

$2C + 6Dt$

When Eq. (3-3) is substituted in Eq. (3-1) and the coefficients of like terms in t on both sides of the resulting equation are equated we obtain $3A + 4B + 2C = [\underline{\hspace{2cm}}]$

- - - - - - - - - - - - - - - - - -

zero

$3B + 8C + 6D = [\underline{\hspace{0.8cm}}]$

- - - - - - - - - - - - - - - - - -

2

$12D + 3C = -3$
$3D = 3$

---

8. If these four equations in the unknowns A, B, C, and D are independent (they are), we should be able to satisfy these equations and make Eq. (3-3) the particular solution of Eq. (3-1).

These values of A, B, C, and D are $A = [\underline{\hspace{1.5cm}}]$; $B = [\underline{\hspace{1.5cm}}]$; $C = [\underline{\hspace{1.5cm}}]$; $D = [\underline{\hspace{1.5cm}}]$.

- - - - - - - - - - - - - - - - -

-38; 12; -5; 1; 3

Then, the particular solution of Eq. (3-1) is

$x = [\underline{\hspace{6cm}}]$.

- - - - - - - - - - - - - - - - -

$x = -38/3 + 12t - 5t^2 + t^3$

If the additional term Et^4 had been included in the proposed particular solution, the value of E could be found to be $[\underline{\hspace{2cm}}]$.

- - - - - - - - - - - - - - - - - -

zero

9. *Generalizing,* then, for inhomogeneous OLDECC in which the *driving function is a polynomial in the independent variable,* the particular solution will also comprise a *polynomial in the independent variable.* The terms in the particular solution will have coefficients which can be determined by substitution of the general form of the solution in the differential equation and equating coefficients of like terms on the two sides of the equation.

Section 2   DETERMINATION OF THE PARTICULAR SOLUTION WHEN DRIVING FUNCTIONS ARE EXPONENTIALS

1. Let us consider the following differential equation:

$$\frac{d^2y}{dx^2} + 4\frac{dy}{dx} + 3y = 12e^{-4x} \tag{3-4}$$

Although this is a different driving function from those already considered, you may by this time have sufficient insight to predict the form of the particular solution.
It is $y = [\underline{\hspace{2cm}}]$.

- - - - - - - - - - - - - - - - - - - -

$$y = Ae^{-4x} \tag{3-5}$$

The proof for the validity of the proposed solution, Eq. (3-5), as a particular solution of (3-6) is made by substituting (3-5) in (3-4).
We have $y = Ae^{-4x}$; $dy/dx = [\underline{\hspace{2cm}}]$; $d^2y/dx^2 = [\underline{\hspace{2cm}}]$.

- - - - - - - - - - - - - - - - - - - -

$-4Ae^{-4x}$ $16Ae^{-4x}$

Hence by substitution of Eq. (3-5) in Eq. (3-4), we obtain $16Ae^{-4x} - 16Ae^{-4x} + 3Ae^{-4x} = 12e^{-4x}$, so that the value of A is $[\underline{\hspace{1cm}}]$.

- - - - - - - - - - - - - - - - - - - -

$A = 4$

Thus the particular solution of (3-4) is $y = [\underline{\hspace{2cm}}]$.

- - - - - - - - - - - - - - - - - - - -

$y = 4e^{-4x}$

---

2. This solution was easy. Before generalizing concerning the solutions for cases in which the driving function is an exponential, let us try another case:

$$\frac{d^2y}{dx^2} + 4\frac{dy}{dx} + 3y = 12e^{-3x} \tag{3-6}$$

Our experience suggests the solution, $y = [\underline{\hspace{2cm}}]$.

- - - - - - - - - - - - - - - - -

$$y = Ae^{-3x} \tag{3-7}$$

for which $dy/dx = [\underline{\hspace{2cm}}]$ and $d^2y/dx^2 = [\underline{\hspace{2cm}}]$.

- - - - - - - - - - - - - - - - -

$-3Ae^{-3x}$ $9Ae^{-3x}$

When these values are substituted in Eq. (3-6), we find $9Ae^{-3x} - 12Ae^{-3x} + 3Ae^{-3x} = 12e^{-3x}$. The value of A required, then, is $[\underline{\hspace{2cm}}]$.

- - - - - - - - - - - - - - - - -

impossible

Hence there is no value of A which will make y = Ae^{-3x} a solution of Eq. (3-6).

3. Perhaps, your previous experience will give you the reason for this apparent inconsistency. The simple particular solution, Eq. (3-7), will not work for Eq. (3-6) because $[\underline{\hspace{4cm}}$
$\underline{\hspace{6cm}}]$.

- - - - - - - - - - - - - - - - -

the coefficient (−3) in the exponent (−3x) is a root of the auxiliary equation $m^2 + 4m + 3 = 0$.

The roots of this auxiliary equation are $m = -3, -1$.
    Your previous experience may also suggest the appropriate particular solution for Eq. (3-6). It is $y = [\underline{\hspace{2cm}}]$.

- - - - - - - - - - - - - - - - -

$$y = Axe^{-3x} \tag{3-8}$$

Reference Equations

3-6 $\dfrac{d^2y}{dx^2} + 4\dfrac{dy}{dx} + 3y = 12e^{-3x}$

3-8 $y = Axe^{-3x}$

---

4. To show that Eq. (3-8) is the particular solution of Eq. (3-6) and find the appropriate value of A, we substitute Eq. (3-8) in Eq. (3-6). We have $y = Axe^{-3x}$.

$dy/dx = [\underline{\hspace{4cm}}]; \; d^2y/dx^2 = [\underline{\hspace{3.5cm}}].$

— — — — — — — — — — — — — — — —

$dy/dx = (e^{-3x} - 3xe^{-3x})A; \; d^2y/dx^2 = (-6e^{-3x} + 9xe^{-3x})A$

The result of substitution of Eq. (3-8) in Eq. (3-6), then, is

$[\underline{\hspace{2.5cm}}] = 12e^{-3x}$

— — — — — — — — — — — — — — — —

$-2Ae^{-3x} = 12e^{-3x}$

---

or $A = -6$; so that the particular solution of Eq. (3-6) is

$y = [\underline{\hspace{2cm}}].$

— — — — — — — — — — — — — — — —

$y = -6xe^{-3x}$

---

5. We have not considered all the possibilities of complications resulting from an exponential driving function in which the coefficient of the independent variable in the exponent is a root of the auxiliary equation. We might have a case in which a double root is involved; for instance,

$$\frac{d^2y}{dt^2} + 4\frac{dy}{dt} + 4y = 16e^{-2t} \tag{3-9}$$

   In this case, our intuition tells us that the particular solution of Eq. (3-9) is probably $y = [\underline{\hspace{2.5cm}}].$

— — — — — — — — — — — — — — — —

$y = At^2\, e^{-2t}$

This is correct, as may readily be shown by substitution; the correct value of $A = [\underline{\hspace{1.5cm}}].$

— — — — — — — — — — — — — — — —

$A = 8$

---

6. We are now ready for a generalization concerning the particular solution of inhomogeneous OLDECC with exponential driving function(s).

Generalization: In a differential equation of the form:

$$\frac{d^n y}{dt^n} + a\, \frac{d^{n-1}y}{dt^{n-1}} + \ldots + ny = Ae^{-k_1 t} + Be^{-k_2 t}$$

The particular solution will comprise terms of the form:
$y = [\underline{\hspace{2cm}}] + [\underline{\hspace{2cm}}].$

- - - - - - - - - - - - - - - -

$y = Ce^{-k_1 t} + De^{-k_2 t}$

unless k_1 and/or k_2 are roots of the *auxiliary equation:*
$[\underline{\hspace{6cm}}].$

- - - - - - - - - - - - - - - -

$m^n + am^{n-1} + \ldots + n = 0$

If k_1 and/or k_2 is a simple root of order n of the auxiliary equation, the particular solution will be of the form:
$y = [\underline{\hspace{2.5cm}}] \text{ and/or } [\underline{\hspace{2cm}}].$

- - - - - - - - - - - - - - - - - - -

$At^{n-1}\, e^{-k_1 t} \text{ and/or } Bt^{n-1}\, e^{-k_2 t}$

7. We now intend to give you some more practice in estimating the trial particular solution under a variety of situations (you need not find the coefficients):

Example 1. (a) $dy/dx + 4y - 2e^x + e^{2x} + 5e^{4x}$
   Your choice for the particular solution is $y = [\underline{\hspace{2.5cm}}].$

- - - - - - - - - - - - - - - -

$y = Ae^x + Be^{2x} + Ce^{4x}$

A, B, and C are as yet undetermined coefficients.
      (b) $dy/dx + 4y = e^x + 2e^{2x} + 3e^{4x} + 7$
The particular solution is of the form: $y = [\underline{\hspace{2.5cm}}].$

- - - - - - - - - - - - - - - -

$y = Ae^x + Be^{2x} + Ce^{4x} + D$

A, B, C, D are definite constants.
      (c) $dy/dx + 2y = 2e^x + 3e^{2x} + 4e^{-2x} + 5e^{4x} + 6$ (3-10)
The particular solution is of the form: $y = [\underline{\hspace{3cm}}$
$\underline{\hspace{2.5cm}}].$

- - - - - - - - - - - - - - -

$y = Ae^x + Be^{2x} + Cxe^{-2x} + De^{4x} + E$ (3-11)

Note that the coefficient of e^{-2x} has been chosen as Cx instead of a pure constant because $m = -2$ is a solution of the auxiliary equation.

> ### Reference Equations
>
> **3-10** $\dfrac{dy}{dx} + 2y = 2e^x + 3e^{2x} + 4e^{-2x} + 5e^{4x} + 6$
>
> **3-11** $Ae^x + Be^{2x} + Cxe^{-2x} + De^{4x} + E$

8. In Case *c* above find the constants *A*, *B*, *C*, *D*, *E* by substitution of Eq. (3-11) in Eq. (3-10) and equating coefficients of like terms in *x*. The result is $A = [\_\_\_]$, $B = [\_\_\_]$, $C = [\_\_\_]$, $D = [\_\_\_]$, $E = [\_\_\_]$.

- - - - - - - - - - - - - - - - - - -

$$A = \frac{2}{3}, \quad B = \frac{3}{4}, \quad C = 4, \quad D = \frac{5}{6}, \quad E = 3$$

If you had difficulty look below:

> EXPLANATION: By substitution: $Ae^x + 2Be^{2x} - 2Cxe^{-2x} + Ce^{-2x} + 4De^{4x} + 2Ae^x + 2Be^{2x} + 2Cxe^{-2x} + 2De^{4x} + 2E = e^x + 3e^{2x} + 4e^{-2x} + 5e^{4x} + 6$.
>
> Equating coefficients (constant parts only) of e^x, e^{2x}, xe^{-2x}, e^{-2x}, and e^{4x}, respectively, we get $A = [\_\_\_]$; $4B = [\_\_\_]$; $2C - 2C = 0$; $C = [\_\_\_]$; $6D = [\_\_\_]$; $2E = 6$.

- - - - - - - - - - - - - - - - - -

$A = 2; \quad 4B = 3; \quad C = 4; \quad 6D = 5; \quad 2E = 6$

> ATTENTION: Do not group together coefficients of e^{-2x} with those of xe^{-2x} by including x inside the balancing equation. Remember that the equations of undetermined coefficients should only involve constants and no variables (dependent or independent).

---

9. *Example 2.* $d^3y/dx^3 + d^2y/dx^2 = x^2 + 2x + e^x$
Your choice for the particular integral is $y = [_____$
$_____]$

- - - - - - - - - - - - - - - - -

$y = Ax^5 + Bx^4 + Cx^3 + Dx^2 + Ex + F + Ge^x$

Note that x^5, x^4, x^3 terms have been included in the general expression for *y* because of the presence in the equation of higher derivatives of *y* with respect to *x*. Careful observation will tell you that $D = [\_\_\_]$; $E = [\_\_\_]$; $F = [\_\_\_]$.

- - - - - - - - - - - - - - - - -

0; 0; 0

10. *Example 3.* $\dfrac{d^4y}{dx^4} + 3\dfrac{d^2y}{dx^2} - 4y = x + \sin x$

Your choice for y = [_____]

- - - - - - - - - - - - - - - - - -

$y = Ax + B + C \sin x + D \cos x$

_____

Find which of the coefficients in the above expression for y are zero. They are [___] = 0; [___] = 0.

- - - - - - - - - - - - - - - - - -

$B = 0$; $D = 0$.

_____

Observation. Since (a) the equation contains only even derivatives of y, (b) the second derivative of $Ax + B$ is zero and (c) the second derivative of $\sin x$ is a mere multiple of $\sin x$, it is clear that B and D are not needed. Thus, $y = Ax + C \sin x$.

   Now, find A and C. They are A = [_____]; C = [_____].

- - - - - - - - - - - - - - - - - -

$A = -1/4$; $C = -1/6$.

_____

11. *Example 4.* $\dfrac{d^2y}{dt^2} + 4y = 3t \cos t$ (3-12)

   The particular integral y = [_____

_____].

- - - - - - - - - - - - - - - - - -

$y = At \cos t + Bt \sin t + C \cos t + D \sin t$ (3-13)

_____

If you did not obtain the particular solution given, see the following explanation.

   EXPLANATION: The complementary function (obtained by set-
   ting the right-hand side = 0) is $y = C_1 \sin 2t + C_2 \cos 2t$. *From*
   t cos t and its derivatives we obtain the terms t cos t, t sin t,
   cos t, sin t. Since the arguments of none of these corresponds to
   roots of the auxiliary equation, we take as a particular solution:
   $y = At \cos t + Bt \sin t + C \cos t + D \sin t$.

> ### Reference Equations
>
> **3-12** $\dfrac{d^2y}{dt^2} + 4y = 3t \cos t$
>
> **3-13** $y = At \cos t + Bt \sin t + C \cos t + D \sin t$

12. Find the actual values of *A, B, C, D* in the particular solution Eq.
(3-13) of Eq. (3-12). They are $A = [\underline{\quad}]$; $B = [\underline{\quad}]$; $C = [\underline{\quad}]$;
$D = [\underline{\quad}]$.

- - - - - - - - - - - - - - - - - -

$A = 1$; $B = 0$; $C = 0$; $D = 2/3$.

Solution: Substitution in the differential equation gives
$3At \cos t + 3Bt \sin t + (2B + 3C) \cos t + (3D - 2A) \sin t = 3t \cos t$
hence $3A = [\underline{\quad}]$, $3B = [\underline{\quad}]$, $2B + 3C = [\underline{\quad}]$, $3D - 2A = [\underline{\quad}]$.

- - - - - - - - - - - - - - - - - -

3, 0, 0, 0.

Be careful to keep the coefficients of $t \cos t$ and $\cos t$, and $t \sin t$
and $\sin t$ separated.

---

13. *Example 5:* To give you extra confidence, we propose that you suggest a form of particular solution for the following equation:

$$\frac{d^2y}{dx^2} - 2\frac{dy}{dx} + 2y = e^x \sin x + e^{-x} \sin 2x + \cos 3x + \cos 4x$$

The particular integral $y = [\underline{\hspace{6cm}}$

$\underline{\hspace{9cm}}]$.

- - - - - - - - - - - - - - - - - -

$y = Ae^x \sin x + Be^x \cos x + Ce^{-x} \sin 2x + De^{-x} \cos 2x + E \cos 3x$
$+ F \sin 3x + Gx \cos 4x + Hx \sin 4x + J \cos 4x + K \sin 4x$

If you are not satisfied, try to evaluate the constants *A* through *K*.
Good luck!

Section 3 SUMMARY OF METHODS OF SOLUTION FOR OLDECC

1. It is appropriate to summarize our findings up to this point. We have dealt both with homogeneous and inhomogeneous OLDECC ordinary linear differential equations with constant coefficients).

   (A) The *homogeneous case* is of the type

   $$a_0 \frac{d^n y}{dx^n} + a_1 \frac{d^{n-1} y}{dx^{n-1}} + \ldots + a_n y = 0 \qquad (3\text{-}14)$$

   The solution is obtained by first substituting, to obtain the auxiliary equation, $y = [\underline{\hspace{1cm}}]$.

   $y = e^{mx}$

   The result is a polynomial equation in m of the following form:
   $a_0 m^n + [\underline{\hspace{2cm}}] + [\underline{\hspace{2cm}}] = 0$

   $$a_0 m^n + a_1 m^{n-1} + \ldots + a_n = 0 \qquad (3\text{-}15)$$

   This is the $[\underline{\hspace{2cm}}]$ equation.

   ---

   auxiliary

....
Reference Equations

3-14 $a_0 \dfrac{d^n y}{dx^n} + a_1 \dfrac{d^{n-1} y}{dx^{n-1}} + \ldots + a_n y = 0$

3-15 $a_0 m^n + a_1 m^{n-1} + \ldots + a_n = 0$
....

2. The next step is to find the roots of the auxiliary equation:

$m = m_1, m_2, m_3 \ldots m_n.$

The solution of the differential equation is, then,

$y = [\text{_____}].$

– – – – – – – – – – – – – – – – – –

$y = k_1 e^{m_1 t} + k_2 e^{m_2 t} + \ldots + k_n e^{m_n t}$

___

unless one or more of the roots of the auxiliary equation are repeated. When repeated roots occur, such that the kth root, for instance, is a quadruple root, the resulting solution of the differential equation will contain corresponding terms of the form:

$y = k_1 e^{m_k x} + k_2 x\, e^{m_k x} + [\text{_____}].$

– – – – – – – – – – – – – – – – – –

$k_3 x^2\, e^{m_k x} + k_4 x^3\, e^{m_k x}$

___

The constants $k_1, k_2, \ldots k_n$ in the homogeneous equation can be determined by which of the following:

  (a) substituting the solution in the differential equation,
  (b) using initial conditions, which are part of the description of the physical situation from which the differential equation arises but are not included in the differential equation.

– – – – – – – – – – – – – – – – – –

(b)

___

3. (*B*) The *inhomogeneous case* is of the type

$$a_0 \frac{d^n y}{dt^n} + a_1 \frac{d^{n-1}y}{dt^{n-1}} \ldots + a_n y = f(t)$$

The function $f(t)$ is termed the [_____] function.

- - - - - - - - - - - - - - - - - - - -

driving or forcing function

The solution of the inhomogeneous equation is the sum of two solutions, termed the [_____] solution and the [_____] solution.

- - - - - - - - - - - - - - - - - - - -

complementary, particular

The complementary solution is the solution of the homogeneous equation obtained by replacing $f(t)$ by [_____].

- - - - - - - - - - - - - - - - - - - -

zero

---

4. (a) This complementary solution is obtained exactly as for a homogeneous equation. It will, for the case of *nonrepeated roots,* be of the form: $y = [$_____$]$.

- - - - - - - - - - - - - - - - - - -

$y = k_1 e^{m_1 t} + k_2 e^{m_2 t} + \ldots$

The number of terms in the complementary solution is necessarily equal to the order of the differential equation. [yes, no]

- - - - - - - - - - - - - - - - - - -

yes

The order of the equation is necessarily equal to the order of the highest derivative that appears in the equation (after any reduction in order which may be possible because of missing terms in the dependent variable). [yes, no]

- - - - - - - - - - - - - - - - - - -

yes

---

5. (*b*) The *particular solution* is a solution of the complete differential equation, i.e., the inhomogeneous equation in which $f(t)$ has *not* been replaced by zero.

The particular solution depends on the form of the driving function. [true, false]

— — — — — — — — — — — — — — — —

true

Particular solutions have been developed up to this point in this work for the following types of driving functions: [⎯⎯⎯⎯⎯⎯ , ⎯⎯⎯⎯⎯⎯ , ⎯⎯⎯⎯⎯⎯ , ⎯⎯⎯⎯⎯⎯⎯] driving functions.

— — — — — — — — — — — — — — — —

constant, sinusoidal, polynomial, exponential

---

6. In each case (except that of the constant driving function) the procedure has been to formulate a trial particular solution with some (general) coefficients.

The actual values of these coefficients have been determined by substituting each proposed trial particular solution in the differential equation and selecting values of the coefficients which will make the proposed solution a valid particular solution of the differential equation, if possible. [true, false]

— — — — — — — — — — — — — — — —

true

Once any valid solution is formed, we know that it is unique, i.e., the only possible particular solution. [yes, no]

— — — — — — — — — — — — — — — —

yes

The procedure used in formulating the trial particular solutions can be formally summarized as in the following.

---

7. RULE I:

The trial function for the evaluation of a *particular integral* (of an OLDECC) is a *linear combination* with *constant undetermined* coefficients of the *forcing function* appearing in the right-hand side of the differential equation and of its *independent derivatives*.

8. RULE II:

If the trial function of Rule I has one or more terms which turn out to correspond to one or more roots of the auxiliary equation, a new trial function must be used which is the *product* of the *trial function of Rule I and the independent variable raised to a power such that a particular solution can be found.*

---

9. The number of terms in the particular solution will always be equal to the order of the differential equation. [yes, no]

----

no

(A simple check of some of our previous work will uncover examples in which the order and number of terms are not equal).

It may not be immediately apparent that these general rules result in the particular solutions previously formulated; however, some careful thought will show that these rules are, in fact, correct.

---

10. In the overall solution of any differential equation, as soon as the particular solution and the complementary solution have been obtained, we are in a position to write the entire solution.

This entire solution will be the sum of the particular solution and the complementary solution.

It will still contain some as yet undetermined coefficients in the [complementary, particular] solution.

----

complementary

The determination of these coefficients requires information not contained in the differential equation. [true, false]

----

true

This information comes from the physical circumstances (problem situations) from which the differential equation arises. [yes, no]

----

yes

Section 4   SUMMARY OF WORK UP TO THIS POINT

At this point, you should know how to

(1) Find the general solution of the homogeneous OLDECC.

(2) Given the initial conditions of the problem, find the values of the coefficients in the homogeneous OLDECC.

from Chapter I

(3) Given an inhomogeneous OLDECC, find the complete solution comprising the complementary solution and the particular solution, for the cases of steady (constant), sinusoidal, exponential, and polynomial driving functions.

(4) Given the initial conditions of the problem, find the value of the coefficients in the complementary solution.

This completes this text in so far as the treatment of the methods of solution of OLDECC is concerned. However, there is a substantial advantage to additional practice with these methods, particularly in relation to some typical problem situations. In Chapter IV, a number of problem situations are described and the corresponding differential equations are developed, with frequent opportunities for the student to participate. We recommend that the student continue through Chapter IV, at least to the point at which he is thoroughly confident that he needs no further practice.

Chapter IV

SOME EXAMPLES OF PROBLEM SOLUTIONS WHICH INVOLVE OLDECC

Section 1   A VARIATION OF AN EARLIER OSCILLATING MASS PROBLEM IN WHICH A DRIVING FORCE HAS BEEN ADDED

1. Section 2 of Part B of Chapter I, Page 32, dealt with a system in which a mass, coupled to a spring, slides back and forth on a smooth plane. The system was set in motion by displacing the mass initially and releasing it. There was no external force applied to the mass after it was released. The mass eventually came to rest at its neutral position.

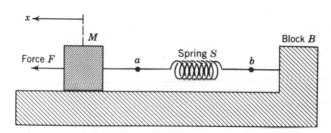

FIGURE 4-1 A physical configuration comprising a mass M, which can slide freely on a smooth plane. The mass is attached to one end, a, of a self-supporting spring, S, the other end of which, b, is attached to a fixed block B. The mass, M, is acted upon by a force, F.

2. In this section, we will consider the same system with a driving force acting on the mass and analyze the resulting motion of the mass. The physical configuration is shown in Fig. 4-1.

A coordinate system for the displacement of the mass and the positive direction of the force acting on the mass are matters of our choice in preparing a mathematical model.

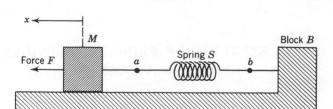

FIGURE 4-1 A physical configuration comprising a mass M, which can slide freely on a smooth plane. The mass is attached to one end, a, of a self-supporting spring, S, the other end of which, b, is attached to a fixed block B. The mass, M, is acted upon by a force, F.

3. The rest position of the mass, i.e., the position of the mass when the force F is zero, will be chosen as $x = 0$ and forces and displacement x from this position will be measured as positive when they act in $+x$ direction.
The forces acting on the mass are
spring force $= [+, -]\ k_s x$ (k_s is force in newtons/meter)

- -

$-$

inertial force $= [+, -]\ M\dfrac{d^2 x}{dt^2}$

- -

$-$

4. We have to decide what to do about frictional forces; for convenience, we will assume that there is a damping force k_d (in newtons/meter/sec) proportional to the velocity of the mass.
The damping force $= [+, -]\ k_d\ [\rule{2cm}{0.4pt}]$.

- - - - - - - - - - - - - - - - - - - -

$-k_d \dfrac{dx}{dt}$

The sum total of all the forces acting on the mass in the above formulation must be zero or $[\rule{4cm}{0.4pt}] = 0$

- - - - - - - - - - - - - - - -

$-M\dfrac{d^2 x}{dt^2} - k_d\dfrac{dx}{dt} - k_s x + F = 0$

or $M\dfrac{d^2 x}{dt^2} + k_d\dfrac{dx}{dt} + k_s x = F$ (4-1)

Reference Equation

4-1 $M \dfrac{d^2x}{dt^2} + k_d \dfrac{dx}{dt} + k_s x = F$

---

5. We will further refine our statement of the problem by adding the following information: (*a*) the mass is to be in its rest position, $x = 0$ at time $t = 0$, the time at which the external force is first applied. The initial conditions are then at $t = 0$. $x = $ [___]; $dx/dt = $ [___].

- - - - - - - - - - - - - - - - - - - -

$x = 0$ (4-2)

$dx/dt = 0$ (4-3)

We will also state that three cases are to be considered:

	Mass	k_d		k_s	
Case A	1 kg.	3	newtons/meter/sec	6.25	newtons/meter
Case B	1 kg.	5	newtons/meter/sec	6.25	newtons/meter
Case C	1 kg.	5.83	newtons/meter/sec	6.25	newtons/meter

---

6. We will start with a fixed driving force of 2 newtons, i.e., $F = 2$.
Remark. The foregoing data amount to a complete description of the physical configuration, reliable only to the extent, however, that such a physical configuration may be accurately described by a model comprising an ideal spring and a frictional force which is proportional to the velocity. Any questions as to the accuracy of this model will not be discussed further in this text. The reader is warned, however, not to assume that any general physical configuration of this description can be accurately described by such a model.

---

7. Our differential equations, then, are

Case A $\dfrac{d^2x}{dt^2} + 3 \dfrac{dx}{dt} + 6.25x = 2$ (4-4)

Case B $\dfrac{d^2x}{dt^2} + 5 \dfrac{dx}{dt} + 6.25x = 2$ (4-5)

Case C $\dfrac{d^2x}{dt^2} + 5.83 \dfrac{dx}{dt} + 6.25x = 2$ (4-6)

Each of these equations gives a *complete* description of the behavior of the corresponding system from time $t = 0$ onward except for the [_____] conditions.

- - - - - - - - - - - - - - - - - - - -

initial

Reference Equations

4-4 $\dfrac{d^2x}{dt^2} + 3\dfrac{dx}{dt} + 6.25x = 2$

4-5 $\dfrac{d^2x}{dt^2} + 5\dfrac{dx}{dt} + 6.25x = 2$

4-6 $\dfrac{d^2x}{dt^2} + 5.83\dfrac{dx}{dt} + 6.25x = 2$

8. The initial conditions to accompany Eqs. (4-4), (4-5), (4-6), as we have already stated, provide additional information (not contained in the differential equations) concerning the physical configuration and must, in general, either be part of the problem statement or deduced from it. In this instance, we made these initial conditions part of the problem statement.

Our initial conditions are at $t = 0$, $x = [\_\_\_]$, $dx/dt = [\_\_\_]$.

– – – – – – – – – – – – – – – – – –

$x = 0$, $dx/dt = 0$

We will now consider the solution of each case in detail.

Case A

9. The differential equation in this case is

$$\frac{d^2x}{dt^2} + 3\frac{dx}{dt} + 6.25x = 2 \qquad (4\text{-}4)$$

The *complementary solution* is found by starting with the homogeneous equation:

$$\frac{d^2x}{dt^2} + 3\frac{dx}{dt} + 6.25x = [\underline{\quad}]$$

- - - - - - - - - - - - - - - - - -

0

___

To obtain the auxiliary equation, we substitute $x = [\underline{\qquad}]$ in this equation

- - - - - - - - - - - - - - - - -

e^{mt}

___

and obtain as the auxiliary equation $[\underline{\qquad\qquad\qquad\qquad}] = 0$

- - - - - - - - - - - - - - - - - - -

$m^2 + 3m + 6.25 = 0$

___

of which the roots are $m = [\underline{\qquad\qquad}], [\underline{\qquad\qquad}].$

- - - - - - - - - - - - - - - - -

$-1.5 + 2i, \ -1.5 - 2i$

___

and the complementary solution, then, is

$$x = e^{-1.5t}(k_1 \cos 2t + k_2 \sin 2t) \qquad (4\text{-}7)$$

Reference Equation

4-7 $x = e^{-1.5t} (k_1 \cos 2t + k_2 \sin 2t$

---

10. Can we proceed immediately to find k_1 and k_2 by substituting the initial conditions in the complementary solution, Eq. (4-7)? [yes, no, explain why].

- - - - - - - - - - - - - - - -

no, the whole solution (complementary plus particular) must be used.

Otherwise this whole solution would not fit the initial conditions. The particular solution is evidently x = [_____].

- - - - - - - - - - - - - - - -

x = 2/6.25 = 0.32

---

and the whole solution, then, without evaluating k_1 and k_2, is
x = [_____].

- - - - - - - - - - - - - - - -

$x = e^{-1.5t} (k_1 \cos 2t + k_2 \sin 2t) + 0.32$ \hfill (4-8)

---

11. To evaluate k_1 and k_2, we substitute the initial conditions at $t = 0$: $x = 0$, $dx/dt = 0$. With these conditions substituted in Eq. (4-8), we find that $k_1 =$ [_____] and $k_2 =$ [_____].

- - - - - - - - - - - - - - - -

$k_1 = -0.32$, $k_2 = -0.24$

---

so that our complete solution is x = [_____
_____].

- - - - - - - - - - - - - - - -

$x = e^{-1.5t} (-0.32 \cos 2t - 0.24 \sin 2t) + 0.32$

The two trigonometric terms can be combined as $A \sin \theta + B \cos \theta = C \sin (\theta + \gamma)$ in which $C =$ [_____] and $\gamma =$ [_____].

- - - - - - - - - - - - - - - -

$C = \sqrt{A^2 + B^2}$, $\gamma = \arctan (B/A)$

---

to obtain the solution in a simpler form:
x = [_____].

- - - - - - - - - - - - - - - -

$x = -0.433 e^{-1.5t} \sin (2t + 56.3^\circ) + 0.32$ \hfill (4-9)

Reference Equation

4-9 $\dot{x} = -0.433e^{-1.5t} \sin(2t + 56.3^{0}) + 0.32$

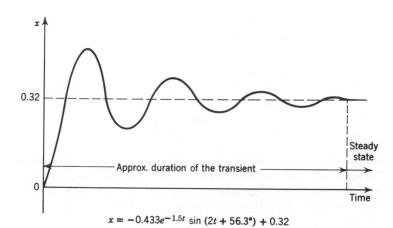

$x = -0.433e^{-1.5t} \sin(2t + 56.3°) + 0.32$

FIGURE 4-2 Sketch of the motion of the mass M of Figure 4-1 under the influence of a constant driving force of *2* kgs. and a damping force of *3* newtons per meter per sec. The solution (4-8) suggests that it actually takes an infinite time for the transient to die out but we would expect, in practice, that static friction (neglected in our solution) would bring the oscillations to a definite stop before an infinite time has elapsed.

12. At this point, we might stop and analyze what this solution, Eq. (4-9), means. It tells us that the mass M oscillates back and forth about its initial position and gradually comes to rest at a new position. [true, false].

----

true

The final resting point is displaced to the left of the initial position by [_____] meters.

0.32 meters

This end rest position is that in which the force of the spring exactly balances the driving force. [false, true]

----

true

We can describe this behavior as characterized by a *transient period,* during which the mass oscillates, but with an amplitude which is ever [increasing, decreasing]

----

decreasing

----

and a *steady-state* condition in which the oscillation has died out and the mass has come to rest. These results are sketched in Fig. 4-2.

Reference Equation

4-9 $x = -0.433e^{-1.5t} \sin (2t + 56.3^\circ) + 0.32$

---

13. The solution given by Eq. (4-9) suggests that it actually takes an infinite time for the transient to die out. [true, false]

– – – – – – – – – – – – – – – – – –

true

In a real situation, it seems probable that static friction (neglected in our solution) would bring the oscillations to a definite stop before an infinite time has elapsed. [true, false]

– – – – – – – – – – – – – – – – – –

true

---

14. The separation of the solution into a "transient" interval and a subsequent steady-state condition is significant. The relation of the complementary and particular solutions to this separation is also notable.

The transient behavior is described by that portion of the solution which we have termed the [_____] solution.

– – – – – – – – – – – – – – – – – –

complementary

The steady-state behavior is described by that portion of the solution which we have termed the [_____] solution.

– – – – – – – – – – – – – – – – – –

particular

If we were interested only in the steady-state behavior, could we neglect the complementary solution? [_____]

– – – – – – – – – – – – – – – – – –

yes

If we were interested only in the transient behavior, could we neglect the particular solution and simply find the complementary solution? [_____]

– – – – – – – – – – – – – – – – – –

no

EXPLANATION: No, even if we are interested in the transient solution only, we could not neglect the particular solution because we cannot find the values of the constants k_1 and k_2 in the complementary solution by using this complementary solution alone. However, we could determine the exponential decay coefficient $e^{-1.5t}$ and the frequency of oscillation $1/\pi$ cycles/sec in the transient solution without any use of the particular solution.

15. Thus in this case, we see that the procedure of separating the solution into two parts, a complementary and a particular solution, not only is convenient from a manipulative point of view but also separates the solution into two important (from a physical point of view) parts, a transient and a steady-state component.

We now turn to the next case, in which the damping term in the differential equation is increased to 5 newtons/meter/sec.

Case B

16. In this case, the situation is described by the differential equation

$$\frac{d^2x}{dt^2} + 5\frac{dx}{dt} + 6.25x = 2 \tag{4-5}$$

The auxiliary equation is [_____] = 0

- - - - - - - - - - - - - - - - - - - -

$m^2 + 5m + 6.25 = 0$

---

and the roots are m = [_____ , _____].

- - - - - - - - - - - - - - - - - - - -

$m = -2.5, -2.5$

Thus the complementary solution is x = [_____]

- - - - - - - - - - - - - - - - - - - -

$x = e^{-2.5t}(k_1 + k_2t)$ $\tag{4-5a}$

---

exactly as in Case B of Chapter I, Part B, Section 3, Pages 38–42, in which there was no driving force.

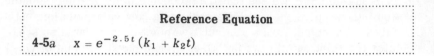

Reference Equation

4-5a $x = e^{-2.5t} (k_1 + k_2 t)$

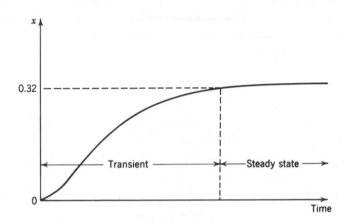

FIGURE 4-3 The position of the mass, M, of Figure 4-1 as a function of time, when the damping is 5 newtons/msec.

17. Are the values of k_1 and k_2 in this case the same as in that of Chapter I, Part B, Section 3, Page 41? [_____]

– – – – – – – – – – – – – – – – – – –

no

EXPLANATION: Since (a) the coefficients k_1 and k_2 are obtained by substituting the initial conditions in the whole solution and (b) the whole solution for the problem now under consideration (which has a driving force) and that of Section 3 of Part B of Chapter I (which had no driving force) are different, we would expect the coefficients k_1 and k_2 to have different values in the two cases.

Is the particular solution in this new case (Case B) the same for Case A? [_____]

– – – – – – – – – – – – – – – – – –

yes

Thus the whole solution is $x = $ [_____].

– – – – – – – – – – – – – – – – –

$x = -e^{-2.5t} (0.32 + 0.8t) + 0.32$ (4-10)

The behavior of the system in this case is sketched in Fig. (4-3).

> **Reference Equation**
>
> **4-10** $x = -e^{-2.5t}(0.32 + 0.8t) + 0.32$

18. In this case, we can also discern a *transient* interval followed by a *steady-state* condition. [yes, no]

— — — — — — — — — — — — — — — — — —

yes

_____

The equation suggests it takes an infinite time for the mass *M* to come to its final position [true, false].

— — — — — — — — — — — — — — — — —

true

_____

however, we actually would expect *static friction* to bring the mass to a stop before infinite time has elapsed.

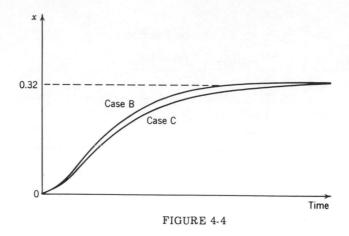

FIGURE 4-4

Case C

19. In this case, the damping term in the differential equation is increased to 5.831 newtons/meter/sec and the differential equation is

$$\frac{d^2x}{dt^2} + 5.831 \frac{dx}{dt} + 6.25x = 2 \qquad (4\text{-}6)$$

The whole solution is $x = [\underline{\hspace{4cm}}] +$
$[\underline{\hspace{3cm}}] + [\underline{\hspace{1.5cm}}]$

– – – – – – – – – – – – – – – – – –

$$x = -0.4715e^{-1.42t} + 0.1515e^{-4.42t} + 0.32 \qquad (4\text{-}11)$$

The motion of the mass is sketched in Fig. 4-4.

The same general conclusions as to *transient* and *steady-state* components apply to the [$\underline{\hspace{3cm}}$] and [$\underline{\hspace{2cm}}$] solutions in this case.

– – – – – – – – – – – – – – – – – –

complementary, particular

Section 2   FURTHER VARIATIONS OF THE EARLIER PROBLEM WITH TIME-VARYING DRIVING FUNCTIONS

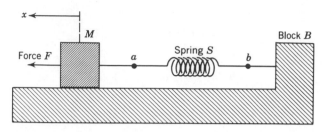

FIGURE 4-1 A physical configuration comprising a mass, M, which can slide freely on a smooth plane. The mass is attached to one end, a, of a self-supporting spring, S, the other end of which, b, is attached to a fixed block B. The mass, M, is acted upon by a force, F.

1. In this section, we will again consider the sliding mass system (Fig. 4-1) but with three different types of driving forces, *each of which is time-varying*. These driving forces will be described by

   Case II: $F = 2 \sin 0.5t$ (sinusoidal driving function)
   Case III: $F = 3 e^{-0.4t}$ (exponential driving function)
   Case IV: $F = 0.5t - 0.25t^2$ (polynomial driving function)

2. The effects of each of these driving forces will be considered for the three previously specified values of the damping constant:

   Case A $k_d = 3$ newtons/meter/sec
   Case B $k_d = 5$ newtons/meter/sec
   Case C $k_d = 5.831$ newtons/meter/sec

   We may summarize the solutions required as those of the following equations.

3. Case   **Equation**

IIA
$$\frac{d^2x}{dt^2} + 3\frac{dx}{dt} + 6.25x = 2\ \sin 0.5t \qquad (4\text{-}12)$$

IIB
$$\frac{d^2x}{dt^2} + 5\frac{dx}{dt} + 6.25x = 2\ \sin 0.5t \qquad (4\text{-}13)$$

IIC
$$\frac{d^2x}{dt^2} + 5.831\frac{dx}{dt} + 6.25x = 2\ \sin 0.5t \qquad (4\text{-}14)$$

IIIA
$$\frac{d^2x}{dt^2} + 3\frac{dx}{dt} + 6.25x = 3\ e^{-0.4t} \qquad (4\text{-}15)$$

IIIB
$$\frac{d^2x}{dt^2} + 5\frac{dx}{dt} + 6.25x = 3\ e^{-0.4t} \qquad (4\text{-}16)$$

IIIC
$$\frac{d^2x}{dt^2} + 5.831\frac{dx}{dt} + 6.25x = 3\ e^{-0.4t} \qquad (4\text{-}17)$$

IVA
$$\frac{d^2x}{dt^2} + 3\frac{dx}{dt} + 6.25x = 0.5t - 0.25t^2 \qquad (4\text{-}18)$$

IVB
$$\frac{d^2x}{dt^2} + 5\frac{dx}{dt} + 6.25x = 0.5t - 0.25t^2 \qquad (4\text{-}19)$$

IVC
$$\frac{d^2x}{dt^2} + 5.831\frac{dx}{dt} + 6.25x = 0.5t - 0.25t^2 \qquad (4\text{-}20)$$

---

4. We expect the student to be able to proceed with the solutions of these differential equations and complete the determination of the behavior of the mass M in each instance by making use of the given initial conditions at $t = 0$ $\begin{cases} x = 0 \\ \dfrac{dx}{dt} = 0 \end{cases}$

However, it may serve to simplify the details of the algebraic work if we can make some generalizations.

---

5. First, we already know from our previous work that the complementary solutions for each value of the damping coefficient, k_d, except for the values of the coefficients k_1 and k_2, are independent of the driving function. [yes, no]

— — — — — — — — — — — — — — — — —

yes

---

6. Next, will we have any identical particular solutions? In the example treated in Section 1 of this chapter, one with a constant driving function, the particular solutions for each of the three equations

$$\frac{d^2x}{dt^2} + 3\frac{dx}{dt} + 6.25x = 2$$

$$\frac{d^2x}{dt^2} + 5\frac{dx}{dt} + 6.25x = 2$$

and

$$\frac{d^2x}{dt^2} + 5.831\frac{dx}{dt} + 6.25x = 2$$

were identical, i.e., x = a constant, 0.32 meters.

---

7. Will the *particular* solutions for each of the equations

$$\frac{d^2x}{dt^2} + 3\frac{dx}{dt} + 6.25x = 2 \sin 0.5t \qquad (4\text{-}12)$$

$$\frac{d^2x}{dt^2} + 5\frac{dx}{dt} + 6.25x = 2 \sin 0.5t \qquad (4\text{-}13)$$

$$\frac{d^2x}{dt^2} + 5.831\frac{dx}{dt} + 6.25x = 2 \sin 0.5t \qquad (4\text{-}14)$$

also be identical? [_____]

- - - - - - - - - - - - - - - - - -

no

> *EXPLANATION: The particular solution for the constant driving function depended only on the term in x in the differential equation, which is the same in all three equations, regardless of the damping. However, when the driving function is a function of time, the particular solution depends upon all terms in the dependent variable and its derivatives in the differential equation. If this conclusion is not apparent to you now, it will be when you go through the steps of finding the solutions.*

---

8. However, except for the numerical values of the coefficients of the functions in the particular solutions of Eqs. (4-12), (4-13), and (4-14) we would expect these particular solutions to be the same. [yes, no]

- - - - - - - - - - - - - - - - - -

yes

Now find the solution of each of the differential equations (4-12) through (4-20). Compare your solutions with the following solutions.

9. (4-12) $x = e^{-1.5t}(0.078 \cos 2t - 0.0195 \sin 2t) + 0.312 \sin 0.5t - 0.078 \cos 0.5t$

(4-13) $x = e^{-2.5t}(0.1184 + 0.154t) + 0.284 \sin 0.5t - 0.1184 \cos 0.5t$

(4-14) $x = 0.148 e^{-1.42t} - 0.017 e^{-4.42t} + 0.27 \sin 0.5t - 0.131 \cos 0.5t$

(4-15) $x = -e^{1.5t}(0.4623 \sin 2t + 0.4777 \cos 2t) + 0.577 \, e^{-0.4t}$

(4-16) $x = e^{-2.5t}(0.483 - 1.21t) + 0.683 \, e^{-0.4t}$

(4-17) $x = -0.95 e^{-1.42t} + 0.22 e^{-4.42t} + 0.735 e^{-0.4t}$

(4-18) $x = e^{-1.5t}(0.04403 \cos 2t - 0.0262 \sin 2t) - 0.04403 + 0.1184t - 0.04t^2$

(4-19) $x = e^{-2.5t}(0.1024 + 0.112t) - 0.1024 - 0.144t - 0.04t^2$

(4-20) $x = 0.1393 \, e^{-1.42t} - 0.0089 \, e^{-4.42t} - 0.1304 + 0.1546t - 0.04t^2$

---

10. We assume that you will by this time have succeeded in finding the correct solutions to these equations.

It is to be noted, in each case, that the complementary solution represents a transient (dying-out) component of behavior of the physical system. [yes, no]

– – – – – – – – – – – – – – – – – –

yes

The particular solution represents a more sustained behavior; in particular, with the periodic (sinusoidal) driving function of (4-12), (4-13), and (4-14), the particular solution represents a steady-state behavior that persists indefinitely after the transient dies out. [yes, no]

– – – – – – – – – – – – – – – – – –

yes

In the other cases, those described by Eqs. (4-15) and (4-20), the driving force either (a) dies out, as in the case of the [*exponential driving function, polynomial driving function*].

– – – – – – – – – – – – – – – – – –

exponential driving function

---

or (b) increases in magnitude indefinitely, as in the case of the [_____].

– – – – – – – – – – – – – – – – – –

polynomial driving function

---

so that the concept of the steady-state solution is less useful.

---

11. After you have followed this work to this point, you should be ready to attack a wide variety of physical problems which can be effectively handled by organized models that yield linear, ordinary differential equations with constant coefficients. This process of problem solving will require much skill in addition to the skill in solving the differential equations developed by this text; remember that no differential equation will give you information that has not fundamentally been written into the equation at the start. The real challenge to your professional skills is going to be in analyzing problems, determining initial conditions, and setting up the differential equation models. The solution of the equation is an orderly and straightforward process in which you must be competent but which will not get you out of trouble if you cannot do the rest of the job. We now continue with more practice examples.

**Section 3 EXAMPLE OF THE SOLUTION OF AN ELECTRIC
CIRCUIT PROBLEM INVOLVING VARIOUS DRIVING FUNCTIONS**

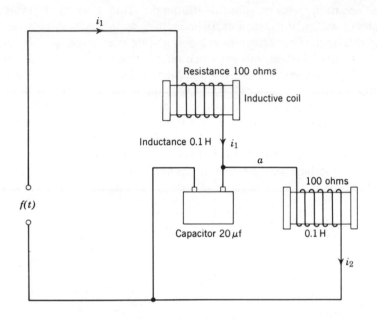

FIGURE 4-5a A circuit comprising two coils and a capacitor.

Let us consider the problem of finding the current i_1 in the circuit of
Fig. 4-5a with voltages $e = f(t)$ as follows:

  Case 1 $e = 10$ volts

  Case 2 $e = 5 \sin 866t$ volts

If you are not familiar with circuit theory skip to Frame 4 on page
112.

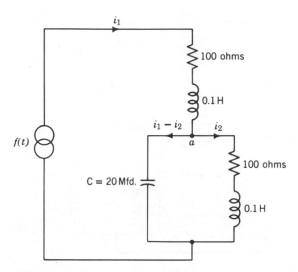

FIGURE 4-5*b* An equivalent circuit of the circuit of Fig. 4-5*a*.

1. To solve this problem, the first step is to find circuit equations, since the usual first step of finding the equivalent circuit of Fig. 4-5*a* has already been done with the result given in Fig. 4-5*b*. The process of finding circuit equations requires the application of [_____] Laws.

— — — — — — — — — — — — — — — — — — —

Kirchhoff's Laws

2. Kirchhoff's Current Law has already been applied at point *a*, Fig. 4-5*b*, so that the current from this point to the left is designated as $i_1 - i_2$. Thus, we need only to apply Kirchhoff's Voltage Law. There are how many independent loops? [one, two, three]

— — — — — — — — — — — — — — — — — — —

two

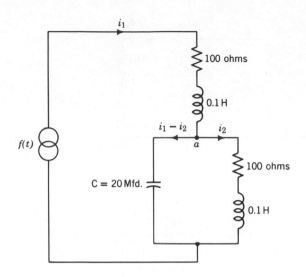

FIGURE 4-5*b* An equivalent circuit of the circuit of Fig. 4-5*a*.

3. The two independent loop equations are

[_____]

[_____]

– – – – – – – – – – – – – – – – – – – –

Any two of the following three loop equations are independent:

$$100_i + 0.1 \frac{di_1}{dt} + 100i_2 + 0.1 \frac{di_2}{dt} = f(t) \qquad (4\text{-}21)$$

$$100i_1 + 0.1 \frac{di}{dt} + \int_0^t \frac{(i_1 - i_2)dt}{20 \times 10^{-6}} + (V_c)_{t=0} = f(t) \qquad (4\text{-}22)$$

$$\int_0^t \frac{(i_1 - i_2)dt}{20 \times 10^{-6}} + (V_c)_{t=0} = 100i_2 + 0.1 \frac{di_2}{dt} \qquad (4\text{-}23)$$

4. We will proceed with a solution which uses Eqs. (4-21) and (4-23). The first step will be [_____

_____].

– – – – – – – – – – – – – – – – – – – –

differentiation of each term in Eq. (4-23) with respect to time.

The result of differentiation of Eq. (4-23) with respect to time is

[_____]

– – – – – – – – – – – – – – – – – –

$$\frac{i_1 - i_2}{20 \times 10^{-6}} = 100 \frac{di_2}{dt} + 0.1 \frac{d^2 i_2}{dt} \quad \text{or} \quad 50{,}000(i_1 - i_2) = 100 \frac{di_2}{dt} +$$

$$0.1 \frac{d^2 i_2}{dt^2} \qquad (4\text{-}24)$$

Reference Equations

4-21 $100i + 0.1\dfrac{di_1}{dt} + 100i_2 + 0.1\dfrac{di_2}{dt} = f(t)$

4-24 $\dfrac{i_1 - i_2}{20 \times 10^{-6}} = 100\dfrac{di_2}{dt} + 0.1\dfrac{d^2i_2}{dt}$ or $50,000\,(i_1 - i_2) =$

$\qquad\qquad 100\dfrac{di_2}{dt} + 0.1\dfrac{d^2i_2}{dt^2}$

---

5. The next step in solving for the current i_1 in the circuit of Fig. 4-5 is to [_____].

- - - - - - - - - - - - - - - - - - - -

solve simultaneously the Eqs. (4-21) and (4-24) for the current i_1.

---

6. There are several methods for finding the single dependent-variable form of two simultaneous differential equations in two dependent variables. One simple method consists of writing the two equations with the operator D, indicating differentiation with respect to time. In this form, Eq. (4-21) becomes

$$(0.1D + 100)i_1 + (0.1D + 100)i_2 = f(t) \qquad (4\text{-}25)$$

Eq. (4-24) becomes

$$50,000i_1 + (-50,000 - 100D - 0.1D^2)i_2 = 0 \qquad (4\text{-}26)$$

The simultaneous solution of these, in determinant form, for i_1 is

$$i_1 = \frac{\begin{vmatrix} f(t) & (0.1D + 100) \\ 0 & (-50,000 - 100D - 0.1D^2) \end{vmatrix}}{\begin{vmatrix} (0.1D + 100) & (0.1D + 100) \\ (50,000) & (-50,000 - 100D - 0.1D^2) \end{vmatrix}}$$

which gives

$$(0.01D^3 + 20D^2 + 20,000D + 10^7)i_1 = (50,000 + 100D + 0.1D^2)\,f(t)$$

Carrying out the indicated differentiation and dividing through by 0.01, we obtain

$$\dfrac{d^3i_1}{dt^3} + 2000\dfrac{d^2i_1}{dt^2} + 2 \times 10^6 \dfrac{di_1}{dt} + 10^9 i_1 = 5\times10^6\,f(t) + 10^4\,f'(t) +$$

$$10f''(t) \qquad\qquad (4\text{-}27)$$

Reference Equation

4-27 $\dfrac{d^3 i_1}{dt^3} + 2000 \dfrac{d^2 i_1}{dt^2} + 2 \times 10^6 \dfrac{di_1}{dt} + 10^9 i_1 = 5 \times 10^6 \, f(t) +$

$10^4 \, f'(t) + 10 f''(t)$

---

7. Before proceeding with the solution we might take a look at some initial conditions. These will ordinarily be part of the problem statement. We have some choice as to these if we are making our own problem statement; the only requirement is that these conditions be physically possible. The conditions chosen will be:

$$\text{at } t = 0, \quad i_1 = 0, \quad V_c = 0, \quad i_2 = 0 \qquad (4\text{-}28)$$

---

8. We remind you that there are to be two cases given for the driving function: Case A $e = 10$ volts; Case B $e = 5 \sin 866t$.
We do not need to separate these cases before finding the complementary solution, so we will proceed to find this first.

   The auxiliary equation for the differential equation (4-27) is
[_____].

- - - - - - - - - - - - - - - - - -

$m^3 + 2000 m^2 + 2 \times 10^6 m + 10^9 = 0$ $\qquad$ (4-29)

---

9. What can we tell about the roots of this equation? There are three of them; hence there must be at least one real root. This root is $m = [$_____$]$.

- - - - - - - - - - - - - - - - - -

$m = -1000$

If you had no trouble finding this root, skip the following explanation and go directly to Frame 11.

---

10. *EXPLANATION for the real root of Eq. (4-29): A third-degree polynomial equation has three roots. Since complex or imaginary roots occur only in pairs, at least one of the three roots must be a real root. This root can be found by the application of the cubic formula or by a "trial" technique such as that of synthetic division. In this case, we used synthetic division. The calculation is shown for $m = -1000$.*

$$1 + 2000 + 2 + 10^6 + 10^9 \quad \underline{|\,-1000}$$
$$\underline{-1000 - 1 \times 10^6 - 10^9}$$
$$1 + 1000 + 1 \times 10^6 - 10^9$$

Thus $m = -1000$ is a root.

Reference Equation

4-29 $m^3 + 2000m^2 + 2 \times 10^6 m + 10^9 = 0$

---

11. Since one root, $m = -1000$, is now known, we can reduce the cubic equation (4-29) to a quadratic equation which contains the remaining two roots. The equation is [_____].

   - - - - - - - - - - - - - - - - -

   $m^2 + 1000m + 1 \times 10^6$ (4-30)

   If your response in Eq. (4-30) was correct, skip the next frame and go to Frame 13.

---

12. Eq. (4-29) is known to have the root $m = -1000$. We can divide Eq. (4-29) by $m + 1000$ to obtain a quadratic equation in the remaining roots.

$$
\begin{array}{r}
m^2 + 1000m\ \ + 1 \times 10^6 \\
m + 1000\ \overline{\big)\ m^3 + 2000m^2 + 2 \times 10^6 m + 1 \times 10^9} \\
\underline{m^3 + 1000m^2} \\
1000m^2 + 2 \times 10^6 m \\
\underline{1000m^2 + 1 \times 10^6 m} \\
1 \times 10^6 m + 1 \times 10^9 \\
\underline{1 \times 10^6 m + 1 \times 10^9} \\
0
\end{array}
$$

   The required quadratic equation, then, is

$$m^2 + 1000m + 1 \times 10^6 \tag{4-30}$$

---

13. The roots of Eq. (4-30) are $m = $ [_____] and $m = $ [_____].

   - - - - - - - - - - - - - - - -

   $m = -500 + i866, \quad m = -500 - i866$

---

14. The complementary solution of Eq. 4-27 without evaluating the coefficients, is, then $i_1 = $ [_____].

   - - - - - - - - - - - - - - - -

   $i_1 = k_1 e^{-1000t} + e^{-500t} (k_2 \cos 866t + k_3 \sin 866t)$ (4-31)

Reference Equation

4-27 $\dfrac{d^3 i_1}{dt^3} + 2000 \dfrac{d^2 i_1}{dt^2} + 2 \times 10^6 \dfrac{di_1}{dt} + 10^9 i_1 = 5 \times 10^6 \, f(t) +$

$10^4 \, f'(t) + 10f''(t)$

---

15. Before the constants k_1, k_2, and k_3 in the homogeneous solution (4-31) can be evaluated, the particular solution for the differential equation (4-27) must be found. We will first consider Case A, for which $f(t) = 10$ volts. The particular solution in this case is
[_____]

- - - - - - - - - - - - - - - - - - - -

$i_1 = \dfrac{5 \times 10^7}{10^9} = 5 \times 10^{-2}$ amperes

---

16. The complete solution of Eq. (4-27), then, for $f(t) = 10$ volts, without evaluating the coefficients in the complementary solution, is
[_____].

- - - - - - - - - - - - - - - - - - - -

$i_1 = k_1 e^{-1000t} + e^{-500t} (k_2 \cos 866t + k_3 \sin 866t) + 5 \times 10^{-2}$

$(4\text{-}32)$

---

17. We need three initial conditions at $t = 0$ to evaluate the constants k_1, k_2, and k_3. These conditions will require values at $t = 0$ of
[_____] [_____] [_____].

- - - - - - - - - - - - - - - - - - - -

i_1, di_1/dt and $d^2 i_1/dt^2$

However, our initial problem statement gave us $t = 0$, $i_1 = 0$, $i_2 = 0$ and $(V_c)_{t=0} = 0$. Only one of these, that for i_1, is one of the required initial conditions. Since only three initial conditions are required and three have been given, it would seem that we should be able to calculate the required conditions from those given. How do we calculate the needed values of di_1/dt and $d^2 i_1/dt^2$?
[_____

_____

_____].

- - - - - - - - - - - - - - - - - - - -

We substitute the three given initial conditions $i_1 = 0$, $i_2 = 0$, and $(V_c)_{t=0} = 0$ at $t = 0$ into the original differential equations to obtain equations from which we can determine values of di_1/dt and $d^2 i_1/dt^2$ at $t = 0$.

18. The original equations, for Case A, were

$$100i_1 + 0.1 \frac{di_1}{dt} + 100i_2 + 0.1 \frac{di_2}{dt} = 10 \tag{4-21}$$

and

$$\int_0^t \frac{(i_1 - i_2)}{20 \times 10^{-6}} dt + (V_c)_{t=0} = 100i_2 + 0.1 \frac{di_2}{dt} \tag{4-23}$$

For $t = 0$ and $i_1 = 0$, $i_2 = 0$ and $(V_c)_{t=0}$, these equations become

$$0.1 \left(\frac{di_1}{dt}\right)_{t=0} + 0.1 \left(\frac{di_2}{dt}\right)_{t=0} = 10$$

$$0.1 \left(\frac{di_2}{dt}\right)_{t=0} = 0 \quad \text{or} \quad \left(\frac{di_2}{dt}\right)_{t=0} = 0$$

Thus we see that $\left(\dfrac{di_1}{dt}\right)_{t=0} = [\underline{\hspace{2cm}}]$.

- - - - - - - - - - - - - - - -

100

---

19. This is the second of the required three conditions. The third can be obtained from differentiated versions of Eqs. (4-21) and (4-23). These are, in order,

$$100 \frac{di_1}{dt} + 0.1 \frac{d^2i_1}{dt^2} + 100 \frac{di_2}{dt} + 0.1 \frac{d^2i_2}{dt^2} = 0 \tag{4-33}$$

$$\frac{i_1 - i_2}{20 \times 10^{-6}} = 100 \frac{di_2}{dt} + 0.1 \frac{d^2i_2}{dt^2} \tag{4-34}$$

Since these equations are valid at all times from $t = 0$ onward, they are valid at $t = 0$.
Substitution of the conditions already given or found for $t = 0$ in Eqs. (4-33) and (4-34) gives

$$10^4 + 0.1 \left(\frac{d^2i_1}{dt^2}\right)_{t=0} + 0.1 \left(\frac{d^2i_2}{dt^2}\right)_{t=0} = 0$$

$$0 = 0.1 \left(\frac{d^2i_2}{dt^2}\right)_{t=0}$$

Hence the value of $\left(\dfrac{d^2i_1}{dt^2}\right)_{t=0} = [\underline{\hspace{2cm}}]$

- - - - - - - - - - - - - - - -

-10^5

> **Reference Equation**
>
> **4-32** $i_1 = k_1 e^{-1000t} + e^{-500t}(k_2 \cos 866t + k_3 \sin 866t) + 5 \times 10^{-2}$

20. Summarized, then, we have three conditions at $t = 0$, $i_1 = 0$, $di_1/dt = 100$, $d^2 i_1/dt^2 = -10^5$ which we will use to determine the constants k_1, k_2, and k_3 in Eq. (4-32). Substituting $t = 0$ and $i_1 = 0$ in Eq. (4-32), we obtain

$$i_1 = k_1 + k_2 + 5 \times 10^{-2} = 0$$

$$k_1 + k_2 = -5 \times 10^{-2} \tag{4-35}$$

Next, we differentiate Eq. (4-32) to obtain di_1/dt. For $(di_1/dt)_{t=0} = 100$ amperes/sec, we obtain the equation

$(di_1/dt)_{t=0} = [\underline{\hspace{6cm}}] = 100.$

- - - - - - - - - - - - - - - - - - -

$-1000k_1 - 500k_2 + 866k_3 = 100$

or

$-10k_1 - 5k_2 + 866k_3 = 1 \tag{4-36}$

Third, we differentiate Eq. (4-32) twice to obtain $d^2 i_1/dt^2$. For $(d^2 i_1/dt^2)_{t=0} = -10^5$, we obtain the equation

$(d^2 i_1/dt^2)_{t=0} = [\underline{\hspace{6cm}}] = -10^5.$

- - - - - - - - - - - - - - - - - -

$= 10^6 k_1 - 5 \times 10^5 k_2 - 8.66 \times 10^5 k_3 = -10^5$

or

$10k_1 - 5k_2 - 8.66k_3 = -1 \tag{4-37}$

21. The three equations (4-35), (4-36), and (4-37) can be solved simultaneously for k_1, k_2, and k_3 to obtain

$[k_1 = \underline{\hspace{2.5cm}}, \quad k_2 = \underline{\hspace{2.5cm}}, \quad k_3 = \underline{\hspace{2.5cm}}].$

- - - - - - - - - - - - - - - - - - -

$k_1 = -5 \times 10^{-2}, \quad k_2 = 0, \quad k_3 = 5.7 \times 10^{-2}$

22. Thus for Case A, the complete solution for the current i_1 is

$$i_1 = -5 \times 10^{-2} \, e^{-1000t} + 5.77 \times 10^{-2} \, e^{-500t} \sin 866t + 5 \times 10^{-2}$$

$$(4\text{-}38)$$

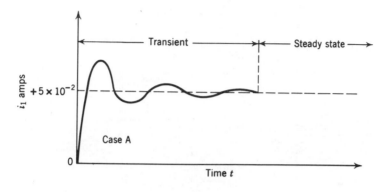

FIGURE 4-6

The behavior of i_1, according to this equation, is sketched in Fig. 4-6. We note, as usual, the presence of a transient component (the complementary solution) and a steady-state component, the particular solution.

Reference Equation

4-31 $i_1 = k_1 e^{-1000t} + e^{-500t}(k_2 \cos 866t + k_3 \sin 866t)$

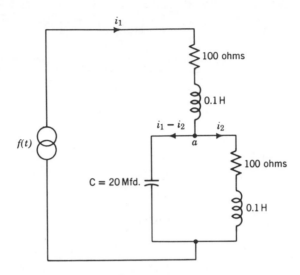

FIGURE 4-5*b* An equivalent circuit of the circuit of Fig. 4-5*a*.

23. We are now ready to proceed with the solution for Case B, in which the driving function, $f(t)$, for the circuit of Fig. 4-5*b* is a voltage, $e = 5 \sin 866t$.

The corresponding differential equation is [_____].

– – – – – – – – – – – – – – – – – –

$$\frac{d^3 i_1}{dt^3} + 2000 \frac{d^2 i_1}{dt^2} + 2 \times 10^6 \frac{di_1}{dt} + 10^9 i_1 = -1.25 \times 10^7 \sin 866t +$$

$$4.33 \times 10^7 \cos 866t \tag{4-39}$$

24. The complementary solution of Eq. (4-39), without evaluating the constants, is $i_1 = $ [_____].

– – – – – – – – – – – – – – – – –

$$k_1 e^{-1000t} + e^{-500t}(k_2 \cos 866t + k_3 \sin 866t) \tag{4-40}$$

This, of course, is the same equation as that for the complementary solution for Case 4, Eq. (4-31). Only the values of the constants k_1, k_2, and k_3 will be different in this case.

Reference Equation

4-39 $\dfrac{d^3 i_1}{dt^3} + 2000\,\dfrac{d^2 i_1}{dt^2} + 2 \times 10^6\,\dfrac{di_1}{dt} + 10^9 i_1 = -1.25 \times$

$$10^7 \sin 866t + 4.33 \times 10^7 \cos 866t$$

25. The particular solution of Eq. (4-39) will be of the form:

$i_1 = [\underline{\hspace{6cm}}]$.

- - - - - - - - - - - - - - - - - -

$i_1 = A \cos 866t + B \sin 866t$ $\hspace{3cm}$ (4-41)

(Did you get this response right? Or did you give

$$i_1 = At \cos 866t + Bt \sin 866t$$

because the argument 866t also appears in the complementary solution.) However, note that the roots of the complementary function are complex:

$$m = -500 + 866i, \ -500 - 866i$$

so that cos 866t and sin 866t do not correspond to roots of the homogeneous equation. In order for these to correspond to the roots of the homogeneous equation, it would be necessary for the roots of the homogeneous equation to be $[m = \underline{\hspace{1.5cm}},\ m = \underline{\hspace{1.5cm}}]$.

- - - - - - - - - - - - - - - - - - -

$m = +866i,\ -866i$

The values of the coefficients A and B in Eq. (4-41) are

$A = [\underline{\hspace{2.5cm}}],\ B = [\underline{\hspace{2.5cm}}]$.

- - - - - - - - - - - - - - - - - -

$A = (-0.00564),\ B = (+0.0373)$

> ### Reference Equation
>
> **4-39** $\dfrac{d^3 i_1}{dt^3} + 2000 \dfrac{d^2 i_1}{dt^2} + 2 \times 10^6 \dfrac{di_1}{dt} + 10^9 i_1 = -1.25 \times$
>
> $\qquad\qquad 10^7 \sin 866t + 4.33 \times 10^7 \cos 866t$

26. Thus the complete solution for the differential equation (4-39), without evaluating the coefficients in the complementary solution, is
$i_1 = k_1 e^{-1000t} + e^{-500t} (k_2 \cos 866t + k_3 \sin 866t) - 0.00566 \cos$
$866t + 0.0373 \sin 866t.$ (4-42)
Can we use the same initial conditions as in Case A at $t = 0$,
$i_1 = 0$, $i_2 = 0$, $(V_c)_{t=0} = 0$? (4-28) [_____]

- - - - - - - - - - - - - - - - - -

yes

_____

These were given to be part of the problem statement. Does this also mean that we can use the same conditions for the derivatives of i_1 derived from the foregoing for (4-28) Case A at $t = 0$, $i_1 = 0$, $di_1/dt = 100$, $d^2 i_1/di^2 = -10^5$? [_____]

- - - - - - - - - - - - - - - - - -

no

_____

We will need to substitute the conditions (4-28) in the original differential equations to determine the values of $(di_1/dt)_{t=0}$ and $(di_2/dt)_{t=0}$.

Reference Equations

4-21 $100i_1 + 0.1 \dfrac{di_1}{dt} + 100i_2 + 0.1 \dfrac{di_2}{dt} = f(t)$

4-23 $\displaystyle\int_0^t \dfrac{(i_1 - i_2)}{20 \times 10^{-6}}\, dt + (V_c)_{t=0} = 100i_2 + 0.1 \dfrac{di_2}{dt}$

4-28 $t = 0, \quad i_1 = 0, \quad i_2 = 0, \quad (V_c)_{t=0} = 0$

27. Since the differential equations for Case B are different from those in Case A, i.e., because the driving function is different, the substitution of the conditions (4-28) for $t = 0$ can be expected to yield different results for $(di_1/dt)_{t=0}$ and $(d^2 i_1/dt^2)_{t=0}$ from these found for Case A.

   For Case B, Eq. (4-21) becomes [_____
_____].

- - - - - - - - - - - - - - - - - - - -

$100i_1 + 0.1 \dfrac{di_1}{dt} + 100i_2 + 0.1 \dfrac{di_2}{dt} = 5 \sin 866t$ \hfill (4-43)

Eq. (4-23) becomes [_____ _____].

- - - - - - - - - - - - - - - - - - - -

$\displaystyle\int_0^t \dfrac{(i_1 - i_2)}{2 \times 10^{-5}}\, dt + (V_c)_{t=0} = 100i_2 + 0.1 \dfrac{di_2}{dt}$ \hfill (4-44)

Actually, Eq. (4-44) is the same as in Case A because the driving function does not appear in either of these equations.

> **Reference Equations**
>
> **4-28** $t = 0,\ i_1 = 0,\ i_2 = 0,\ (V_c)_{t=0} = 0$
>
> **4-42** $i_1 = k_1 e^{-1000t} + e^{-500t}\ (k_2 \cos 866t + k_3 \sin 866t) - 0.00566 \cos 866t + 0.0373 \sin 866t.$
>
> **4-43** $100 i_1 + 0.1 \dfrac{di_1}{dt} + 100 i_2 + 0.1 \dfrac{di_2}{dt} = 5 \sin 866t$
>
> **4-44** $\displaystyle\int_0^t \dfrac{(i_1 - i_2)}{2 \times 10^{-5}}\, dt + (V_c)_{t=0} = 100 i_2 + 0.1 \dfrac{di_2}{dt}$

28. We proceed now to set $t = 0$ and substitute the conditions (4-28). The result from (4-43) is

$$0.1 \left(\frac{di_1}{dt}\right)_{t=0} + 0.1 \left(\frac{di_2}{dt}\right)_{t=0} = 0$$

and from (4-44)

$$0 = 0.1 \left(\frac{di_2}{dt}\right)_{t=0}$$

This gives us: $(di_1/dt)_{t=0} = [\underline{\hspace{1cm}}]$.

- - - - - - - - - - - - - - - - - - -

zero

Next, we differentiate Eqs. (4-43) and (4-44) to obtain

$$100 \frac{di_1}{dt} + 0.1 \frac{d^2 i_1}{dt^2} + 100 \frac{di_2}{dt} + 0.1 \frac{d^2 i_2}{dt^2} = 4330 \cos 866t$$

$$\frac{i_1 - i_2}{2 \times 10^{-5}} = 100 \frac{di_2}{dt} + 0.1 \frac{d^2 i_2}{dt^2}$$

These give us $(d^2 i_1/dt^2)_{t=0} = [\underline{\hspace{1cm}}]$.

- - - - - - - - - - - - - - - - - - -

4.33×10^4

---

29. The initial conditions for Case B, then, are $t = 0$, $i_1 = 0$ amperes, $di_1/dt = 0$ amperes per sec, $d^2 i_1/dt^2 = 4.33 \times 10^4$ amperes per sec^2. Substituting these initial conditions in the complete solution (4-42), we obtain $k_1 = [\underline{\hspace{1cm}}]$, $k_2 = [\underline{\hspace{1cm}}]$, $k_3 = [\underline{\hspace{1cm}}]$.

- - - - - - - - - - - - - - - - - - -

$k_1 = 0.01241,\ k_2 = -0.00675,\ k_3 = -0.0351$

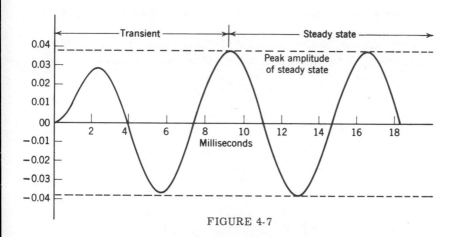

FIGURE 4-7

30. The complete solution for Case B then, is

$$i_1 = 0.01241 \ e^{-1000t} + e^{-500t} (-0.00675 \cos 866t - 0.0351 \sin 866t)$$
$$- 0.00566 \cos 866t + 0.0373 \sin 866t \qquad (4\text{-}45)$$

This solution is sketched in Fig. 4-7. We note, as usual, the presence of a transient component (the complementary solution) and a steady-state component, the particular solution.

Section 4 SOME SIMPLE EXAMPLES FOR FINAL PRACTICE

The following are some simple physical situations on which you may wish to practice. These are electric circuit problems, for which the equivalent circuits are shown in Fig. 4-8.

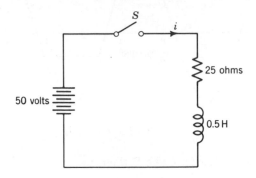

FIGURE 4-8a Switch S closed at $t = 0$. $(i)_{t=0}$

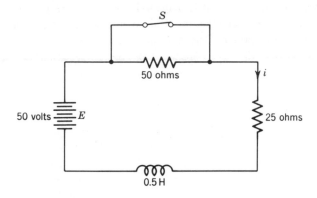

FIGURE 4-8b Switch S opened at $t = 0$. $(i)_{t=0} = 2$ amperes

If you are not familiar with electric circuit theory, you can skip the indicated step of writing the circuit equations and proceed to the step of solving the differential equation, which does not require the circuit theory.

1. First, referring to the circuit of Fig. 4-8a, the differential equation
   for the current i is

   $t \geq 0$ [_____].
   $t \geq 0$
   $- - - - - - - - - - - - - - - - - -$

   $$0.5 \frac{di_1}{dt} + 25\,i = 50 \tag{4-46}$$

   Remark. Since this is a first-order differential equation, it could be
   solved by separation of the variables without the application of the
   methods developed in this book. However, these methods are appli-
   cable and their use in this instance will give you practice.

Reference Equation

4-46 $0.5 \dfrac{di_1}{dt} + 25\,i = 50$

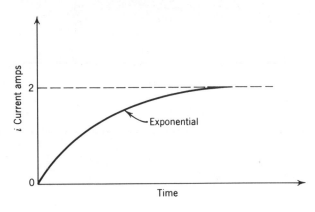

FIGURE 4-9a

2. The auxiliary equation corresponding to the differential Eq. (4-46) is [_____].

$0.5m + 25 = 0$

The root is $m = $ [_____].

$m = -50$

The particular solution is $i = $ [_____].

$i = 2$

The complete solution, without evaluating the constant in the complementary solution, is $i = $ [_____].

$i = ke^{-50t} + 2$ (4-47)

Using the given initial condition $(i)_{t=0} = 0$ we can evaluate the constant k. This constant is $k = $ [____].

$k = -2$

The complete solution is $i = $ [_____].

$i = 2\,(1 - e^{-50t})$ amperes   (4-48)

The solution is sketched in Fig. 4-9a.

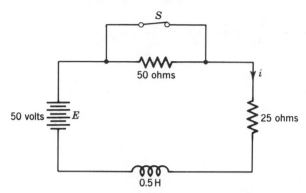

FIGURE 4-8*b* Switch *S* opened at $t = 0$. $(i)_{t=0} = 2$ amperes

3. The differential equation for *i* in the circuit of Fig. 4-8*b* is
$t \geq 0$ [_____].

– – – – – – – – – – – – – – – – – –

$$0.5\frac{di}{dt} + 75i = 50 \qquad\qquad\qquad (4\text{-}49)$$

Reference Equation

4-49 $0.5 \dfrac{di}{dt} + 75i = 50$

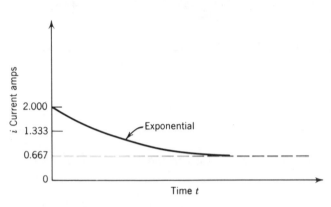

FIGURE 4-9*b*

4. The auxiliary equation corresponding to the differential equation
(4-49) is $i = [$_____$]$.

$0.5m + 75 = 0$

The root of the auxiliary equation is $m = [$_____$]$.

$m = -150$

The particular solution is $i = [$_____$]$.

$i = 0.667$

The complete solution, without evaluating the constant in the complementary solution, is $i = [$_____$]$.

$i = ke^{-150t} + 0.667$

We can evaluate the constant k, using the initial conditions given
with the equivalent circuit. The constant k is $k = [$_____$]$ amperes.

1.333

The complete solution is $i = [$_____$]$.

$i = 1.333\, e^{-150t} + 0.667$ amperes

The solution is sketched in Fig. 4-9*b*.

Section 5   CLOSING REMARKS

By this time, if the student is still with us, he should have acquired a formidable skill in obtaining the solutions of given OLDECC. Of course, most of the cases given have involved equations of no higher order than the second. The reason for this limitation has been the algebraic complexity of finding the roots of an algebraic polynomial equation, here, always the auxiliary equation, when the equation is of higher than the second degree (except for a few simple cases such as the biquadratic case). The authors do not mean to imply that solutions of higher-order OLDECC cannot be found; in fact, when the coefficients in the OLDECC have numerical values, the solutions can, in principle, always be found. If numerical values are used, however, a "general" solution is not obtained. The usefulness of a solution for particular values of the configuration parameters is somewhat limited. When a design in which many choices are possible is under consideration, it is usually necessary to know what may happen when the configuration parameters vary. This would seem to require a process of solving the OLDECC again and again, each time with a different numerical value of the parameters, with values spaced over the range on which the designer needs to study the behavior of the problem configuration. In today's engineering world, it is becoming increasingly common to use in higher than second-order cases an analog computer, a hybrid computer, or an analog simulation program with a digital computer to perform the actual computation. The solution of first- and second-order problems in *general* forms by the methods of this text remains an important tool, however.

The student should not be overconfident concerning the impact of a high degree of skill in solving OLDECC (if he has acquired such either through the use of this text or in other ways) on his ability to solve engineering or physical science problems which lead to first- or second-order OLDECC. The most important step in problem solution remains that of setting up the differential equation model of the problem configuration. A skillful solution of an invalid differential equation model remains an invalid solution.

Finally, the student is reminded that an index is provided for this text. If at some time in the future, as a result of disuse, he finds himself at a loss with respect to some feature of the OLDECC solution, he may be able to use the index to go back to some intermediate starting point which will refresh his mind on the particular point without reworking the complete text.

ANSWERS TO TEST OF MINIMUM MATHEMATICAL
PREREQUISITES REQUIRED

1. (a) $3.4 + 6.2x + 0.03x^2 - 0.4x^3$

   (b) $b + 3cx^2 - 5dx^4$

   (c) $a\omega_1 \cos \omega_1 t - b\omega_2 \sin \omega_2 t + g \sin \omega_3 t + g\omega_3 t \cos \omega_3 t$

   (d) ke^{ky}

2. (a) $m = -1.25$

   (b) $m = 1, \ m = 3$

   (c) $m = 3, \ m = 3$

   (d) $m = 1 + i, \ m = 1 - i, \ i = \sqrt{-1}$
        or
        $m = 1 + j, \ m = 1 - j, \ j = \sqrt{-1}$

   (e) $m = 2, \ m = 3$

3. (a) $A = 3.4, \ B = 4.1$

   (b) $A = 1.1, \ B = 0.2$

   (c) $A = +10, \ B = +5$

4. (a) a

   (b) kt^2

   (c) i/c

5. $A = -9, \ B = 12$

6. $\dfrac{d^2z}{dx^2} + 2\dfrac{dx}{dx} + 8z = 0$

7. (a) $\dfrac{e^{ix} + e^{-ix}}{2}$

   (b) $\dfrac{e^{ix} - e^{-ix}}{2i}$ or $-i\left(\dfrac{e^{-ix} - e^{-ix}}{2}\right)$

8. (a) $y = 5 \cos 6x - 5i \sin 6x$

   (b) $A(\cos t + i \sin t)$

   (c) $\cos kt + i \sin kt$

INDEX